T0202835

Communications
in Computer and Information Science 1584

More information about this series at https://link.springer.com/bookseries/7899

Patricia Pesado · Gustavo Gil (Eds.)

Computer Science – CACIC 2021

27th Argentine Congress, CACIC 2021
Salta, Argentina, October 4–8, 2021
Revised Selected Papers

 Springer

Editors
Patricia Pesado (iD)
National University of La Plata
La Plata, Argentina

Gustavo Gil
National University of Salta
Salta, Argentina

ISSN 1865-0929   ISSN 1865-0937 (electronic)
Communications in Computer and Information Science
ISBN 978-3-031-05902-5   ISBN 978-3-031-05903-2 (eBook)
https://doi.org/10.1007/978-3-031-05903-2

This Springer imprint is published by the registered company Springer Nature Switzerland AG
The registered company address is: Gewerbestrasse 11, 6330 Cham, Switzerland

Preface

Welcome to the selected papers of the 27th Argentine Congress of Computer Science (CACIC 2021), held in an interactive, live online setting due to the COVID-19 situation during October 4–8, 2021. CACIC 2021 was organized by the National University of Salta (UNSa) on behalf of the Network of National Universities with Computer Science Degrees (RedUNCI).

CACIC is an annual congress dedicated to the promotion and advancement of all aspects of computer science. Its aim is to provide a forum within which the development of computer science as an academic discipline with industrial applications is promoted, trying to extend the frontier of both the state of the art and the state of the practice. The main audience for, and participants of, CACIC are seen as researchers in academic departments, laboratories, and industrial software organizations.

CACIC 2021 covered the following topics: intelligent agents and systems; distributed and parallel processing; computer technology applied to education; graphic computation, visualization and image processing; databases and data mining; software engineering; hardware architectures, networks and operating systems; innovation in software systems; signal processing and real-time systems; innovation in computer science education; computer security; digital governance; and smart cities.

This year, the congress received 130 submissions. Each submission was reviewed by at least two, and on average 3.1, Program Committee members and/or external reviewers. A total of 76 full papers, involving 316 different authors from 50 universities, were accepted. According to the recommendations of the reviewers, 21 of these papers were selected for this book.

CACIC 2021 incorporated a number of special activities, including a plenary lecture, two discussion panels, a special track on Digital Governance and Smart Cities, and an International School with four courses.

Special thanks go to the members of the different committees for their support and collaboration. Also, we would like to thank the local Organizing Committee, reviewers, lecturers, speakers, authors, and all conference attendees. Finally, we want to thank Springer for their support of this publication.

April 2022

Patricia Pesado
Gustavo Gil

Organization

The 27th Argentine Congress of Computer Science (CACIC 2021) was organized by the National University of Salta (UNSa) on behalf of the Network of National Universities with Computer Science Degrees (RedUNCI).

Program Chairs

Patricia Pesado (RedUNCI Chair)	National University of La Plata, Argentina
Gustavo Daniel Gil	National University of Salta, Argentina

Editorial Assistant

Pablo Thomas	National University of La Plata, Argentina

Program Committee

Maria Jose Abásolo	National University of La Plata, Argentina
Claudio Aciti	National University of Central Buenos Aires, Argentina
Hugo Alfonso	National University of La Pampa, Argentina
Jorge Ardenghi	National University of the South, Argentina
Marcelo Arroyo	National University of Río Cuarto, Argentina
Hernan Astudillo	Federico Santa María Technical University, Chile
Sandra Baldasarri	University of Zaragoza, Spain
Javier Balladini	National University of Comahue, Argentina
Luis Soares Barbosa	University of Minho, Portugal
Rodolfo Bertone	National University of La Plata, Argentina
Oscar Bria	National University of La Plata, Argentina
Nieves R. Brisaboa	University of La Coruña, Spain
Carlos Buckle	National University of Patagonia San Juan Bosco, Argentina
Alberto Cañas	University of West Florida, USA
Ana Casali	National University of Rosario, Argentina
Silvia Castro	National University of the South, Argentina
Alejandra Cechich	National University of Comahue, Argentina
Edgar Chávez	Michoacana University of Saint Nicolás de Hidalgo, Mexico
Carlos Coello Coello	CINVESTAV, Mexico

Uriel Cuckierman	National Technological University, Argentina
Armando E. De Giusti	National University of La Plata, Argentina
Laura De Giusti	National University of La Plata, Argentina
Marcelo De Vincenzi	Inter-American Open University, Argentina
Claudia Deco	National University of Rosario, Argentina
Beatriz Depetris	National University of Tierra del Fuego, Argentina
Javier Díaz	National University of La Plata, Argentina
Juergen Dix	TU Clausthal, Germany
Ramón Doallo	University of La Coruña, Spain
Domingo Docampo	University of Vigo, Spain
Jozo Dujmovic	San Francisco State University, USA
Marcelo Estayno	National University of Lomas de Zamora, Argentina
Elsa Estevez	National University of the South, Argentina
Jorge Eterovic	National University of La Matanza, Argentina
Marcelo A. Falappa	National University of the South, Argentina
Pablo Rubén Fillotrani	National University of the South, Argentina
Jorge Finocchieto	CAECE University, Argentina
Daniel Fridlender	National University of Cordoba, Argentina
Fernando Emmanuel Frati	National University of Chilecito, Argentina
Carlos Garcia Garino	National University of Cuyo, Argentina
Luis Javier García Villalba	Complutense University of Madrid, Spain
Marcela Genero	University of Castilla-La Mancha, Spain
Sergio Alejandro Gómez	National University of the South, Argentina
Eduard Groller	Vienna University of Technology, Austria
Roberto Guerrero	National University of San Luis, Argentina
Jorge Ierache	National University of Buenos Aires, Argentina
Tomasz Janowski	Danube University Krems, Austria
Horacio Kuna	National University of Misiones, Argentina
Laura Lanzarini	National University of La Plata, Argentina
Guillermo Leguizamón	National University of San Luis, Argentina
Fernando Lopez Gil	University of Zaragoza, Spain
Ronald Prescott Loui	Washington University in St. Louis, USA
Emilio Luque Fadón	Autonomous University of Barcelona, Spain
Maria Cristina Madoz	National University of La Plata, Argentina
Maria Alejandra Malberti	National University of San Juan, Argentina
Cristina Manresa Yee	University of the Balearic Islands, Spain
Javier Marco	University of Zaragoza, Spain
Mauricio Marin	National University of Santiago de Chile, Chile
Ramon Mas Sanso	University of the Balearic Islands, Spain
Orlando Micolini	National University of Cordoba, Argentina

Sponsors

RedUNCI

Network of Universities with Degrees in Computer Science

National University of Salta

Contents

Intelligent Agents and Systems

Internal Behavior Analysis of SA Using Adaptive Strategies to Set
the Markov Chain Length 3
  Carlos Bermudez, Hugo Alfonso, Gabriela Minetti, and Carolina Salto

Distributed and Parallel Processing

Performance Comparison of Python Translators for a Multi-threaded
CPU-Bound Application 21
  Andrés Milla and Enzo Rucci

Computer Technology Applied to Education

Administrative Modular Integral System Participatory. Design in Software
Development 41
  *Verónica K. Pagnoni, Diego F. Craig, Juan P. Méndez,
  and Eduardo E. Mendoza*

Systematic Review of Educational Methodologies Implemented During
the COVID-19 Pandemic in Higher Education in Ibero-America 49
  Omar Spandre, Paula Dieser, and Cecilia Sanz

Software Tool for Thematic Evolution Analysis of Scientific Publications
in Spanish 64
  Santiago Bianco, Laura Lanzarini, and Alejandra Zangara

Higher Education and Virtuality from an Inclusion Approach 78
  *Javier Díaz, Ivana Harari, Ana Paola Amadeo, Alejandra Schiavoni,
  Soledad Gómez, and Alejandra Osorio*

Graphic Computation, Images and Visualization

Automatic Extraction of Heat Maps and Goal Instances of a Basketball
Game Using Video Processing 95
  *Gerónimo Eberle, Jimena Bourlot, César Martínez,
  and Enrique M. Albornoz*

xii   Contents

Software Engineering

A Sound and Correct Formalism to Specify, Verify and Synthesize
Behavior in BIG DATA Systems 109
  Fernando Asteasuain and Luciana Rodriguez Caldeira

Data Variety Modeling: A Case of Contextual Diversity Identification
from a Bottom-up Perspective 124
  Líam Osycka, Agustina Buccella, and Alejandra Cechich

Best Practices for Requirements Validation Process 139
  Sonia R. Santana, Leandro R. Antonelli, and Pablo J. Thomas

Databases and Data Mining

Distribution Analysis of Postal Mail in Argentina Using Process Mining 159
  Victor Martinez, Laura Lanzarini, and Franco Ronchetti

Anorexia Detection: A Comprehensive Review of Different Methods 170
  *María Paula Villegas, Leticia Cecilia Cagnina,
  and Marcelo Luis Errecalde*

GPU Permutation Index: Good Trade-Off Between Efficiency and Results
Quality 183
  Mariela Lopresti, Fabiana Piccoli, and Nora Reyes

A Comparison of DBMSs for Mobile Devices 201
  *Fernando Tesone, Pablo Thomas, Luciano Marrero, Verena Olsowy,
  and Patricia Pesado*

Hardware Architectures, Networks, and Operating Systems

Service Proxy for a Distributed Virtualization System 219
  Pablo Pessolani, Marcelo Taborda, and Franco Perino

Innovation in Software Systems

Ontology Metrics and Evolution in the GF Framework for Ontology-Based
Data Access 237
  Sergio Alejandro Gómez and Pablo Rubén Fillottrani

Ground Segment Anomaly Detection Using Gaussian Mixture Model
and Rolling Means in a Power Satellite Subsystem 254
  Pablo Soligo, Germán Merkel, and Ierache Jorge

Signal Processing and Real-Time Systems

Prototype of a Biomechanical MIDI Controller for Use in Virtual
Synthesizers 269
 Fernando Andrés Ares, Matías Presso, and Claudio Aciti

Computer Security

Introduction of Metrics for Blockchain 285
 Javier Díaz, Mónica D. Tugnarelli, Mauro F. Fornaroli, Lucas Barboza,
 Facundo Miño, and Juan I. Carubia Grieco

Digital Governance and Smart Cities

Data Quality Applied to Open Databases: "COVID-19 Cases"
and "COVID-19 Vaccines" 297
 Ariel Pasini, Juan Ignacio Torres, Silvia Esponda, and Patricia Pesado

Lexical Analysis Using Regular Expressions for Information Retrieval
from a Legal Corpus 312
 Osvaldo Mario Spositto, Julio César Bossero, Edgardo Javier Moreno,
 Viviana Alejandra Ledesma, and Lorena Romina Matteo

Author Index 325

Intelligent Agents and Systems

Internal Behavior Analysis of SA Using Adaptive Strategies to Set the Markov Chain Length

Carlos Bermudez[1], Hugo Alfonso[1], Gabriela Minetti[1(✉)],
and Carolina Salto[1,2]

[1] Facultad de Ingeniería, Universidad Nacional de La Pampa, General Pico,
Argentina
{bermudezc,alfonsoh,minettig,saltoc}@ing.unlpam.edu.ar
[2] CONICET, General Pico, Argentina

Abstract. In the context of Simulated Annealing (SA) algorithms, a
Markov chain transition corresponds to a move in the solution space.
The number of transitions or the Markov chain (MC) length at each
temperature step is usually constant and empirically set. However, adaptive methods to compute the MC length can be used. This work focus
on the effect of using different strategies to set the MC length in the
SA behavior. To carry out this analysis, the Water Distribution Network
Design (WDND) problem is selected, since it is a multimodal and NP-hard problem interesting to optimize. The results indicate that the use
of adaptive strategies to set the MC length improves the solution quality
versus the static one. Moreover, the proposed SA achieves the scalability
property when the WDND solution space size is considered.

Keywords: Simulated annealing · Markov chain length · Water
distribution network design · Optimization

1 Introduction

The development of new and efficient algorithms remains an interesting research
area when solving complex problems, mainly those considered in the complexity
theory (NP-complete or NP-hard) [1]. The Simulated Annealing algorithm [2,3]
is specially applied to solve NP-hard problems where it is difficult to find the
optimal solution or even near-to-optimum solutions [4,5]. The SA simulates the
energy changes in a system subjected to a cooling process until it converges to
an equilibrium state (steady frozen state); in this context the material states
correspond to problem solutions, the energy of a state to a solution cost, and
the temperature to a control parameter.

The Simulated Annealing introduces two elements. The first one is the
Metropolis algorithm [6] in which some states that do not improve energy are
accepted when they allow the algorithm to explore more the possible solution
space. The second element decreases the temperature, again by analogy with

P. Pesado and G. Gil (Eds.): CACIC 2021, CCIS 1584, pp. 3–17, 2022.
https://doi.org/10.1007/978-3-031-05903-2_1

metal annealing. The SA makes a certain number of Metropolis decisions for each temperature value, obtaining a Markov chain of a certain length. In this way, the SA consists of two cycles: one external for temperatures and the other internal, named Metropolis. Most SA literature proposals use a static Markov chain (MC) length in the Metropolis cycle for each temperature [7]. But adaptive strategies to dynamically establish each MC length for the SA algorithm are also present in the literature [8,9].

The main contribution of our research is to enlarge the knowledge concerning the MC length influence on the efficiency and efficacy of a SA when solving optimization problems. In particular, we tackle the Water Distribution Network Design (WDND), which was defined as a multi-period, single-objective, and gravity-fed design optimization problem [10]. A hybrid SA (HSA), presented in [11], was used as a starting point to consider the different strategies to compute the MC length. We conduct experiments by applying HSA with distinct configurations on publicly available [12] and real-world [13] instances of the WDND problem. Furthermore, we analyze and compare these results considering the published ones in the literature. The present work extends the one presented in CACIC 2021 [14], by including high-dimensional instances of the WDND problem in the experimentation together with an in-depth explanation of the analysis results. Consequently, this has an impact on the statistical studies. Moreover, these new results are compared with the ones obtained by the literature proposal. Accordingly, four research questions (RQs) arise:

- $RQ1$: Can the adaptive strategies modify or improve the HSA performance in contrast with the static one?
- $RQ2$: How does HSA perform relative to WDND solvers in the literature when the problem scales in complexity?
- $RQ3$: How do variable MC lengths affect the HSA behavior?
- $RQ4$: How is the correlation between the solution quality and MC lengths for the strategy that leads to a good trade-off between quality and computational effort?

The article organization is summarized as follows. Section 2 describes of the SA algorithm, whereas Sect. 3 addresses the strategies for computing MC lengths. Then, the experimental design and the methodology used are explained in Sect. 4. The analysis and comparison of the HSA behavior when solving the WDND problems are presented in Sect. 5. Finally, we summarize our most important conclusions and sketch out our future work.

2   Simulated Annealing

The Simulated Annealing [2] is an optimization method which explores through the solution space using a stochastic hill-climbing process. The more efficient SA formulations are based on two cycles: one external for temperatures and other internal, named Metropolis. The same Markov chain length in the Metropolis cycle are usually used for each temperature T (a control parameter with $T > 0$).

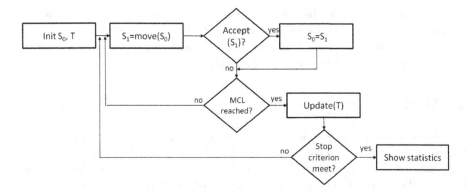

Fig. 1. Scheme of the SA algorithm.

The SA algorithm can be seen like a sequence of Markov chains, where each Markov chain is constructed for descending values of T.

Figure 1 shows the general scheme of a SA algorithm. The SA begins with the initialization of the temperature, T, and the generation of a feasible initial solution, S_0, for the target problem. After that two overlapping cycles begin. A new trial solution, S_1, is obtained by applying a move to the current solution, S_0, to explore other areas of the search space. At this point, S_1 is accepted with the Boltzmann probability, as shown in Eq. 1 for a minimization problem:

$$P = \begin{cases} 1 & \text{if } f(S_1) \leqslant f(S_0) \\ e^{-(f(S_1)-f(S_0))/T} & \text{otherwise} \end{cases} \tag{1}$$

This process generates a Markov chain, which is repeated for a number of steps, named as Markov chain length. After that, the temperature in the SA algorithm is sequentially lowered until the system freezes by a cooling schedule. Finally, the SA ends the search when the total evaluation number or the temperature equilibrium $(T = 0)$ is achieved.

Consequently, the SA's process contains two cycles:

- The external cycle slowly reduces the temperature to decrease defects, thus minimizing the system energy.
- The internal cycle, named Metropolis [6], generates a new potential solution (or neighbor of the current state) to the considered problem by modifying the current solution, according to a predefined criterion. For each temperature value, the Markov chain length usually remains without changes in the Metropolys cycle.

The search space exploration is strengthened when the T is high. But at low temperatures, the algorithm only exploits a promising region of the solution space, intensifying the search. The annealing procedure involves taking enough steps at each temperature (internal cycle). The number of steps aims to keep the system close to equilibrium until the system approaches the ground state. Traditionally, the equilibrium can be achieved by maintaining the temperature constant for a limited number of iterations; but adaptive strategies can be considered. The main objective of this work is to identify how sensitive the SA can be to the number of these iterations, by considering different strategies to compute the Markov chain length to solve NP-hard problems.

3   Markov Chain Length

The Simulated Annealing constructs a sequence of temperatures T_1, T_2, and thus following. At each step of this sequence, the SA does a set of moves to neighboring positions. If a move increases the solution quality, the change is accepted; but if the solution quality decreases, the change is rejected with some probability, in which case the SA simply keeps the old solution at that step. Such a stochastic sequence construction is called a Markov chain, and the number of moves k is called Markov chain length. If the MC length is long, the candidate solution will visit many points in the search space. There are few research articles in the literature concerning the effect of the MCL on the solution quality and annealing speed [8,9]. Different strategies to compute the MCL can be considered. For instance, the MCL can be determined experimentally and considered static throughout the search (MCLs), but also MCL can set adaptively depending on the optimization function variation. For this last strategy, which depend on the characteristics of the search, we consider two different alternatives: MCLa1 and MCLa2. In what follows, a description of each one is carried out.

MCLs. The static strategy assumes that each T value is held constant for a fixed number of iterations, defined before the search starts. In this work, each T value remains constant for 30 iterations, a widely used number in the scientific community.

MCLa1. Cardoso et al. [8] consider that the equilibrium state is not necessarily attained at each temperature. Here, the cooling schedule is applied as soon as an improved candidate (neighbor) solution is generated. In this way, the computational effort can be reduced without compromising the solution quality.

MCLa2. This strategy, proposed by Ali et al. [9], uses both the worst and the best solutions found in the Markov chain (inner loop) to compute the next MCL. MCLa2 increases the number of function evaluations at a given temperature if the difference between the worst and the best solutions increases. But if an improved solution is found, the MCL remains unchanged.

4   Experimental Design

This section describes the experimental design carried out to analyze the behavior of the SA introduced in [11], named HSA, using different MCL strategies to solve the WDND problem. The upcoming subsections briefly describe the target test problem, the WDND instances used for the experimentation, followed by the methodology and the parameters considered.

4.1 Multi-period Water Distribution Network Design

The optimal design of Water Distribution Networks (WDNs) consists in finding the least-cost pipe configuration for a given WDN topology (node placement and network connectivity). The design of WDNs gives rise to an optimization problem called the water distribution network design optimization problem. The aim is to find the least cost design, in terms of optimal pipe types, that satisfies hydraulic laws and customer requirements. Consequently, the decision variables are the diameters for each pipe in the network. In this work, the WDND problem is defined as simple-objective, multi-period, and gravity-fed. Two restrictions are considered: the limit of water speed in each pipe and the demand pattern that varies in time. The network can be modeled by a connected graph, which is described by a set of nodes $N = \{n_1, n_2, ...\}$, a set of pipes $P = \{p_1, p_2, ...\}$, a set of loops $L = \{l_1, l_2, ...\}$, and a set of commercially available pipe types $T = \{t_1, t_2, ...\}$. The objective of the WDND problem is to minimize the Total Investment Cost (TIC) in a water distribution network design. The TIC value is obtained by the formula shown in Eq. 2.

$$\min TIC = \sum_{p \in P} \sum_{t \in T} L_p IC_t x_{p,t} \qquad (2)$$

where IC_t is the cost of a pipe p of type t, L_p is the length of the pipe, and $x_{p,t}$ is the binary decision variable that determines whether the pipe p is of type t or not. The objective function is constrained by: physical laws of mass and energy conservation, minimum pressure demand in the nodes, and the maximum speed in the pipes, for each time $\tau \in T$ (see [13] for more details).

4.2 HydroGen Networks

The HydroGen networks [12] arise from different WDNs, named HG-MP-i. In our previous work [14], ten different networks were considered. The present article extension enlarges the number of networks by including 15 new ones of high-dimensionality, consequently HG-MP-i goes from 1 to 25. Table 1 presents the features of each HG-MP-i network, regarding meshedness coefficient (MC), number of pipes (m), demand nodes (d), and water reservoir (WR). The demand nodes are divided into five categories (domestic, industrial, energy, public services, and commercial demand nodes), with their corresponding baseload and demand pattern. In this way, each HG-MP-i network consists of five different instances, totaling 125 instances. Each HG-MP-i uses 16 different pipe types summarized in Table 2, including their corresponding roughness and their unit costs (expressed in US dollars). The combination of 16 pipe types and the number of pipes of each instance determines the network dimensionality. Moreover, the solution space size is determined by the number of pipes m and the number of possible pipe types t, hence the number of possible solutions is t^m. For example, the solution space is $16^{100} = 2.6E+120$ for the smallest and lowest dimensional HG-MP-1 network, and 16^{906} is the solution space size for the biggest one, the HG-MP-25 network. Therefore, nine different categories are obtained to know

Table 1. Information on the WDND networks.

Network	MC	m	d	WR	Network	MC	m	d	WR
HG-MP-1	0.20	100	73	1	HG-MP-14	0.15	499	385	3
HG-MP-2	0.15	100	78	1	HG-MP-15	0.10	495	413	3
HG-MP-3	0.10	99	83	1	HG-MP-16	0.20	606	431	4
HG-MP-4	0.20	198	143	1	HG-MP-17	0.15	607	465	4
HG-MP-5	0.15	200	155	1	HG-MP-18	0.10	608	503	5
HG-MP-6	0.10	198	166	1	HG-MP-19	0.20	708	503	5
HG-MP-7	0.20	299	215	2	HG-MP-20	0.15	703	538	5
HG-MP-8	0.15	300	232	2	HG-MP-21	0.10	707	586	5
HG-MP-9	0.10	295	247	2	HG-MP-22	0.20	805	572	6
HG-MP-10	0.20	397	285	2	HG-MP-23	0.15	804	615	6
HG-MP-11	0.15	399	308	2	HG-MP-24	0.10	808	669	6
HG-MP-12	0.10	395	330	3	HG-MP-25	0.20	906	644	6
HG-MP-13	0.20	498	357	2	GP-Z2-2020	0.20	282	222	1

Table 2. Available pipe types and their corresponding costs for HydroGen networks.

Number	Diam. (mm)	Roughness	Cost	Number	Diameter	Roughness	Cost
1	20	130	15	9	200	130	116
2	30	130	20	10	250	130	150
3	40	130	25	11	300	130	201
4	50	130	30	12	350	130	246
5	60	130	35	13	400	130	290
6	80	130	38	14	500	130	351
7	100	130	50	15	600	130	528
8	150	130	61	16	1,000	130	628

HG-MP1-3, HG-MP4-6, HG-MP7-9, HG-MP10-12, HG-MP13-15, HG-MP16-18, HG-MP19-21, HG-MP22-24, and HG-MP25 regarding the number of pipes and the size of the solution space of each instance, and showing an incremental network complexity.

4.3   GP-Z2-2020: A Real Network

The GP-Z2-2020 network [13] is composed of 222 domestic demand nodes and only one water reservoir. Moreover, this zone is connected with three other ones through some peripherical nodes which have different demand patterns. Table 3 summarizes available pipe diameters in the local market, their corresponding roughness, and their unit costs (expressed in US dollars). The area is residential with demand according to the current distribution of the customers in 584 plots but considering a development pattern over a timespan of 30 years. The daily pattern demand corresponds to the summer period (based on the model demand of historical records) having a maximum resolution of one hour. The total number

Table 3. Pipe types and their corresponding costs for the GP-Z2-2020 network.

Number	Diam. (mm)	Roughness	Cost	Number	Diam. (mm)	Roughness	Cost
1	63	110	2.85	5	315	110	69.10
2	90	110	5.90	6	400	110	110.89
3	110	110	8.79	7	450	110	140.15
4	125	110	11.00	8	630	110	273.28

of possible combinations of design for a set of 8 commercial pipe types and 282 pipes is $8^{282} = 4.7E+254$ (the size of the solution space) that makes the instance in a difficult case to solve; this shows the importance of optimization.

4.4 Methodology and Experimental Setup

This subsection presents the HSA parametric setting to solve the WDND problem, using a test set of growing complexity, to answer the RQs formulated in Sect. 1. HSA uses the three MCL strategies presented in Sect. 3: the static (MCLs) and the two adaptive (MCLa1 and MCLa2) ones, arising three HSA configurations. HSA employs the random cooling scheme [15] and a seed temperature set in 100 (see [16] for a justification of this parameter selection). Furthermore, this hybrid SA uses the EPANET 2.0 toolkit [17] to solve the hydraulic equations since this hydraulic solver is applied in most existing works. The stop condition is set to 1,500,000 EPANET calls to ensure a fair empirical comparison between the HSA results and the obtained by the literature algorithms. Finally, the reader can find details about the WDND solution representation and its operators in our previous work [13].

Since we deal with stochastic algorithms, we have performed 30 independent runs per WDND instance and for each HSA configuration. We have carried out a statistical analysis of the results that consists of the following steps. Before performing the statistical tests, we first check whether the data follow a normal distribution by applying the Shapiro-Wilks test. Where the data are distributed normally, we later apply an ANOVA test. Otherwise, we use the Kruskal-Wallis (KW) test. These statistical studies allow us to assess whether or not there are meaningful differences between the compared algorithms with $\alpha = 0.05$. These pairwise algorithm differences are determined by carrying out a post hoc test, as the Tukey test for ANOVA or the Wilcoxon test for the KW one.

5 HSA Result Analysis

The result analysis is carried out, considering the performance and internal behavior of the proposed HSA configurations (MCLs, MCLa1, and MCLa2). Firstly, we study the HSA performance from three points of view: solution quality, computational effort, and comparison against the results reported in the literature. Then, we analyze the HSA internal behavior by studying the correlation between MC lengths, TIC values, and temperatures. Finally, we examine the MCLa1 strategy effect on the HSA to solve different complexity networks.

Table 4. The best known TIC values found by our proposals and ILS.

Network	MCLs	MCLa1	MCLa2	ILS	Network	MCLs	MCLa1	MCLa2	ILS
HG-MP-1	**298000**	**298000**	**298000**	**298000**	HG-MP-14	880125	875925	**874694**	889000
HG-MP-2	245330	245330	245330	**245000**	HG-MP-15	1007089	**991666**	1000304	1018000
HG-MP-3	310899	310706	**310493**	318000	HG-MP-16	1265160	**1241866**	1247640	1273000
HG-MP-4	592048	**590837**	592036	598000	HG-MP-17	1398753	1367711	**1363682**	1400000
HG-MP-5	**631000**	**631000**	**631000**	**631000**	HG-MP-18	1170698	1150994	**1146374**	1167000
HG-MP-6	617821	**609752**	614917	618000	HG-MP-19	1742546	1692950	**1663580**	1761000
HG-MP-7	648372	644568	**639932**	653000	HG-MP-20	1763737	1759724	1734885	**1732000**
HG-MP-8	795996	792436	**790037**	807000	HG-MP-21	1810225	1729166	**1719957**	1807000
HG-MP-9	716944	715863	**712450**	725000	HG-MP-22	2262652	2189742	**2173312**	2320000
HG-MP-10	730916	**712847**	727818	724000	HG-MP-23	2461917	**2331700**	2346248	2466000
HG-MP-11	859032	859032	**838539**	856000	HG-MP-24	2351756	**2253724**	2277259	2300000
HG-MP-12	909890	912042	**905239**	920000	HG-MP-25	2857251	**2678023**	2699468	2839000
HG-MP-13	1070981	**1057307**	1086745	1127000	GP-Z2-2020	355756	366684	358717	**347596**
Summarizing						2/26	11/26	14/26	5/26

5.1   HSA Performance

To analyze the HSA performance, we consider the better solutions found by each configuration, the effort required by the search, and the comparison of HSA results against the ILS ones [18], a well-known WDND solver in the literature.

The minimum TIC values for the HSA considering the three MCL strategies reported in Table 4 allow us to study the HSA performance from the solution quality point of view. The last row summarizes the number of networks where each configuration obtains the best-known TIC value of all networks. To complete this study, we use the relative distance between the best-known TIC value and the best TIC value of each HSA configuration as the error metric. In the following this metric is called TIC error value. The Figs. 2 and 3.a) show the distribution of the HSA errors grouped by the HydroGen and GP-Z2-2020 networks and MCL strategies. Boxplots with different colors show statistically different behaviors. From these results, we observe that adaptive HSA configurations improve the static one in 22 of 26 networks. For HG-MP-i with $i \in [7, 25]$ and GP-Z2-2020 networks, the MCLs behavior is statistically different (gray boxplot) to both MCLa1 and MCLa2, which keep similar behavior for all networks (white boxplots). From these results, we can infer that the adaptive MCL strategies allow enhancing the HSA performance versus the static one, answering the first RQ positively.

The required time to execute the whole search process supports the analysis of the HSA performance from the computational effort point of view. Figures 4 and 3.b) show the distribution of this metric grouped by HydroGen and GP-Z2-2020 networks and MCL strategies. First, we observe that the HSA runtimes grow as the instance complexity increases for all configurations, an expected result. Second, MCLs is the quickest strategy for all networks, whereas adaptive HSA configurations increment significantly the total runtime. However, the MCLa1 runtimes are significantly less than the required ones by MCLa2 for HG-MP-i with $i \in [1, 4]$ networks. However, this difference is short for the remain-

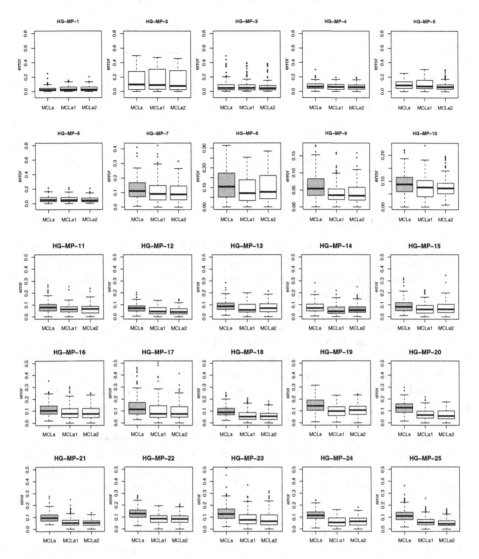

Fig. 2. BoxPlots of TIC error values found by HSA and each MCL strategies for HydroGen networks.

ing networks, but still, MCLa1 presents a slight advantage against MCLa2 in average.

Finally, the comparison of HSA results against the ILS ones is carried out and reported in Table 4. We detect that the three HSA configurations improve the ILS performance, finding better average TIC values than ILS for 21 of 26 networks and the same results in another two. Therefore, the $RQ2$ is affirmatively answered, becoming the HSA into a robust solver for the WDND problem.

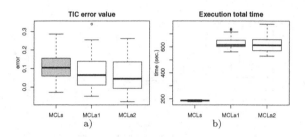

Fig. 3. BoxPlots of TIC error values, a), and total time, b), required by HSA and each MCL strategies for WDND real network (GP-Z2-2020).

5.2 HSA Internal Behavior

To answer how the MCL strategies affect the HSA performance, we study the relationship between the solution quality and the MC lengths and the temperature schedules. We also analyze if the relationship between the solution quality and the MC lengths varies taking into account the network complexity. Figures 5, 6, and 7 show the upper triangular matrix of scatter plots, where the correlation between the three variables: TIC error values, MC lengths, and temperatures are graphically presented, for each HSA configuration and all networks. The Spearman's correlation coefficient, R, is calculated in every comparison and measures the linear correlation between two data sets. R belongs to the range $[-1, 1]$ and expresses the type of association between two variables. If $R > 0$ indicates a positive relationship between the two variables (as values of one variable increase, values of the other variable also increase). When $R < 0$ indicates a negative relationship (as values of one variable increase, values of the other variable decrease). A $R = 0$ means that no linear correlation exists between the variables. However, correlation coefficients like Spearman assume a linear relationship between variables. Even if $R = 0$, a non-linear relationship might exist.

The MCLs strategy maintains constant (equal to 30) the MC length during the whole search process. Consequently, no linear correlation exists between this length and the solution quality ($R = 0$), as Fig. 5 shown. Instead, the temperature reduction is related to the solution quality because the network TIC error values decrease during the annealing process ($R = 0.48$).

As we explain in Sect. 3, the adaptive strategies calculate on runtime the MC length according to the optimization function variation. The lengths computed by MCLa1 and MCLa2 vary in the range $[730, 6690]$ (see histograms of Figs. 6 and 7), becoming a factor that impacts positively in the solution quality (R=0.1 and R=0.06, respectively). As Figs. 6 and 7 show, this impact is different when

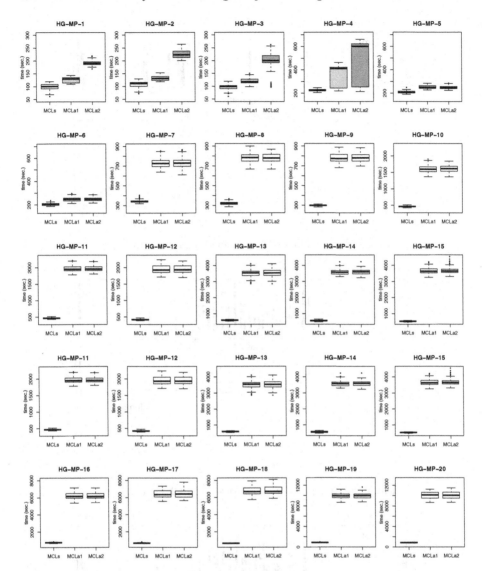

Fig. 4. BoxPlots of the total time (in seconds) required by HSA and each MCL strategies to solve the HydroGen networks.

the first adaptive strategy is used, because only MCLa1 enable to decrease the MC length. Consequently, when HSA uses MCLa1 can reduce the temperature more times during the search process. This situation allows reaching a better equilibrium between exploration and exploitation in the search space, leading to a solution quality improvement.

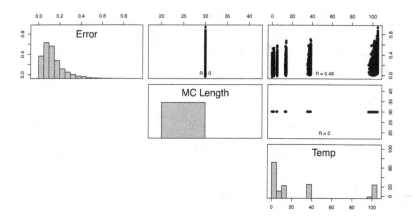

Fig. 5. Scatter plots of correlation for MCLs strategy.

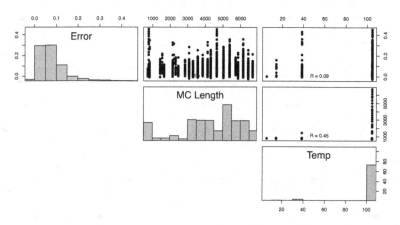

Fig. 6. Scatter plots of correlation for MCLa1 strategy.

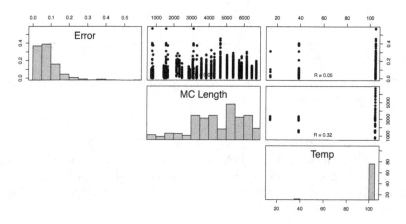

Fig. 7. Scatter plots of correlation for MCLa2 strategy.

Table 5. The Spearman's coefficient, R, and the MC length range for MCLa1 strategy in each network category.

Networks	R	MC length	
		Minimum	Maximum
HG-MP1-3	0.11	730	6690
HG-MP4-6	−0.09	1430	1660
HG-MP7-9	−0.29	2150	2470
HG-MP10-12	−0.22	2850	3300
HG-MP13-15	−0.06	3570	4130
HG-MP16-18	−0.11	4310	5030
HG-MP19-21	−0.36	5030	5860
HG-MP22-24	−0.17	5720	6690

Furthermore, we analyze the temperature behavior in more detail. As we can deduce in the above paragraph, the variability of the MC length also affects the temperature schedule. MCLa1 and MCLa2 maintains a positive correlation (R = 0.45 and R = 0.32, respectively), indicating that a diminution in the lengths is associated with a temperature reduction. According to this analysis, the $RQ3$ is also satisfactorily answered because HSA modifies its behavior with the MC length variability.

Finally, MCLa1 provides the best tradeoff between solution quality and computational effort. Moreover, if the focus is on the equilibrium between exploitation and exploration, MCLa1 also is the MCL strategy with better behavior.

5.3 HSA Internal Behavior Using MCLa1

Given MCLa1 strategy enable to decrease the MC length, this section is devoted to study, in more detail, the MCLa1 effect on the HSA to solve different complexity networks, considering the categories defined in Sect. 4.2. In this sense, Table 5 shows Spearman's correlation coefficient, between the TIC error values and the MC lengths, and the minimum and maximum MC lengths for each network category for this HSA configuration. Therefore, the following analysis answers the $RQ4$.

As we explain in Sect. 4.2, each category groups networks with similar complexities and solution space sizes, which grow by the increments in the number of pipes. This growth in the complexity causes variations in the correlation between the TIC error values and the MC lengths, as the second column in Table 5 shows. On the one hand, the range of MC lengths is much wider for the smallest networks grouped in the HG-MP1-3 category than for the remaining ones. Besides, for this category, the number of EPANET calls is lesser because of the solution size, as a consequence, the Metropolis cycle is executed in more opportunities, contributing to enlarge the range of MC lengths. These particularities lead to a positive R coefficient, implying that both variables increase or decrease together. On the other hand, when the network complexity augments, the negative R coefficient indicates that the TIC error values decrease if the MC length increase.

6 Conclusions

In this work, we study the Hybrid Simulated Annealing behavior using different MCL strategies to solve high-dimensional instances of the WDND problem. The SA algorithms usually employ static Markov chain lengths in the Metropolis cycle for each temperature value. However, adaptive strategies to compute this length, which depend on the optimization function variability, are often underutilized. The motivation of this research was to analyze if the HSA performance changes depending on the network dimensionality. Accordingly, we tested diverse solution space sizes from 16^{100} to 16^{906}. We consider 126 different WDND problem instances solved by HSA, in which MCL is computed by static (MCLs) and two adaptive (MCLa1 and MCLa2) strategies.

The experimentation results allowed us to infer that, for all solution space sizes, the adaptive MCL strategies improve the HSA solution quality versus the MCLs and the reported ones in the literature, answering the two first RQs. In other words, we empirically and statistically verified that no significant changes in the HSA performance are observed when the solution space size scales. Furthermore, the last two RQs could be answered from the internal behavior analysis point of view. We deduce that the variability in the MC length modifies the HSA behavior positively. Although, the correlation kind between the length and the solution quality for the smallest networks (HG-MP1-3 category) differs from the remaining ones.

As future research lines, we are interested in enhancing the HSA runtime for the high-dimensional networks by proposing a parallel HSA. Furthermore, we plan to use big-data frameworks to implement our PHSA solver, which will allow us to deal with high-dimensional WDND problems.

Acknowledgements. The authors acknowledge the support of Universidad Nacional de La Pampa (Project FI-CD-107/20) and the Incentive Program from MINCyT, Argentina. The last two authors are grateful for the support of the HUMAN-CENTERED SMART MOBILITY (HUMOVE) project, PID2020-116727RB-I00, Spain. The last author is also funded by CONICET, Argentina.

References

1. Garey, M.R., Johnson, D.S.: Computers and Intractability; A Guide to the Theory of NP-Completeness. W. H. Freeman & Co., USA (1990)
2. Kirkpatrick, S., Gelatt, C.D., Jr., Vecchi, M.: Optimization by simulated annealing. Science **220**, 671–680 (1983)
3. Černý, V.: Thermodynamical approach to the traveling salesman problem: an efficient simulation algorithm. J. Optim. Theory Appl. **45**, 41–51 (1985)

4. Borysenko, O., Byshkin, M.: Coolmomentum: a method for stochastic optimization by langevin dynamics with simulated annealing. Sci. Rep. **11**, 10705 (2021)
5. Lin, S.-W., Cheng, C.-Y., Pourhejazy, P., Ying, K.-C., Lee, C.-H.: New benchmark algorithm for hybrid flowshop scheduling with identical machines. Expert Syst. Appl. **183**, 115422 (2021)
6. Metropolis, N., Rosenbluth, A.W.R.M.N., Teller, A.H.: Nonequilibrium simulated annealing: a faster approach to combinatorial minimization. J. Chem. Phys. **21**, 1087–1092 (1953)
7. Alba, E., Blum, C., Isasi, P., León, C., Gómez, J.A. (eds.), Frontmatter. Wiley, Hoboken (2009). https://doi.org/10.1002/9780470411353
8. Cardoso, M., Salcedo, R., de Azevedo, S.: Nonequilibrium simulated annealing: a faster approach to combinatorial minimization. Ind. Eng. Chem. Res. **33**, 1908–1918 (1994)
9. Ali, M., Törn, A., Viitanen, S.: A direct search variant of the simulated annealing algorithm for optimization involving continuous variables. Comput. Oper. Res. **29**(1), 87–102 (2002)
10. Cunha, M., Sousa, J.: Water distribution network design optimization: simulated annealing approach. J. Water Resour. Plan. Manage. **125**, 215–221 (1999)
11. Alfonso, H., Bermudez, C., Minetti, G., Salto, C.: A real case of multi-period water distribution network design solved by a hybrid SA. In: XXVI Congreso Argentino de Ciencias de la Computación (CACIC), pp. 21–3 (2020). http://sedici.unlp.edu.ar/handle/10915/113258
12. De Corte, A., Sörensen, K.: Hydrogen. http://antor.uantwerpen.be/hydrogen. Accessed 27 Jun 2018
13. Bermudez, C., Alfonso, H., Minetti, G., Salto, C.: Hybrid simulated annealing to optimize the water distribution network design: a real case. In: Pesado, P., Eterovic, J. (eds.) CACIC 2020. CCIS, vol. 1409, pp. 19–34. Springer, Cham (2021). https://doi.org/10.1007/978-3-030-75836-3_2
14. Bermudez, C., Alfonso, H., Minetti, G.F., Salto, C.: Performance analysis of simulated annealing using adaptive markov chain length. In: XXVII Congreso Argentino de Ciencias de la Computación (CACIC 2021), pp. 21–30 (2021)
15. Bermudez, C., Minetti, G., Salto, C.: SA to optimize the multi-period water distribution network design. In: XXIX Congreso Argentino de Ciencias de la Computación (CACIC 2018), pp. 12–21 (2018)
16. Bermudez, C., Salto, C., Minetti, G.: Designing a multi-period water distribution network with a hybrid simulated annealing. In: XLVIII JAIIO: XX Simposio Argentino de Inteligencia Artificial (ASAI 2019), pp. 39–52 (2019)
17. Rossman, L.: The EPANET Programmer's Toolkit for Analysis of Water Distribution Systems (1999)
18. De Corte, A., Sörensen, K.: An iterated local search algorithm for water distribution network design optimization. Network **67**(3), 187–198 (2016)

Distributed and Parallel Processing

Performance Comparison of Python Translators for a Multi-threaded CPU-Bound Application

Andrés Milla[1] and Enzo Rucci[2]([✉])

[1] Facultad de Informática, UNLP, 1900 La Plata, Buenos Aires, Argentina
[2] III-LIDI, Facultad de Informática, UNLP - CIC,
1900 La Plata, Buenos Aires, Argentina
erucci@lidi.info.unlp.edu.ar

Abstract. Currently, Python is one of the most widely used languages in various application areas. However, it has limitations when it comes to optimizing and parallelizing applications due to the nature of its official CPython interpreter, especially for CPU-bound applications. To solve this problem, several alternative translators have emerged, each with a different approach and its own cost-performance ratio. Due to the absence of comparative studies, we have carried out a performance comparison of these translators using *N-Body* as a case study (a well-known problem with high computational demand). The results obtained show that CPython and PyPy presented poor performance due to their limitations when it comes to parallelizing algorithms; while Numba and Cython achieved significantly higher performance, proving to be viable options to speed up numerical algorithms.

Keywords: Numba · Cython · N-body · CPU-bound · Parallel computing

1 Introduction

Since it came out in the early 1990s, Python has now become one of the most popular languages [19]. Still, Python is considered "slow" compared to compiled languages like C, C++, and Fortran, especially for CPU-bound applications[1]. Among the causes of its poor performance are its nature as an interpreted language and its limitations when implementing multi-threaded solutions [15]. In particular, its main problem is the use of a component called *Global Interpreter Lock* (GIL) in the official CPython interpreter. GIL only allows executing a single thread at a time, which leads to sequential execution. To overcome this limitation, processes are usually used instead of threads, but this comes at the cost of higher consumption of resources and a higher programming cost due to having a distributed address space [10].

---

[1] Programs that perform a large number of calculations using the CPU exhaustively.

P. Pesado and G. Gil (Eds.): CACIC 2021, CCIS 1584, pp. 21–38, 2022.
https://doi.org/10.1007/978-3-031-05903-2_2

Even though there are alternative interpreters to CPython, some of them have the same problem, as in the case of PyPy [8]. In the opposite sense, some interpreters do not use GIL at all, such as Jython [4]. Unfortunately, Jython uses a deprecated version of Python [4,9], which limits future support for programs and the ability to take advantage of features provided by later versions of the language. Other translators allow the programmer to disable this component, as in the case of Numba, a JIT compiler that translates Python into optimized machine code [6]. Numba uses a Python feature known as decorators [1], to interfere as little as possible in the programmer's code. Finally, Cython is a static compiler that allows transpiling[2] Python codes to the equivalent C ones, and then compiling it to object code [2]. It also allows disabling GIL and using C libraries, such as OpenMP [3], which is extremely useful for developing multi-threaded programs.

When implementing a Python application, a translator must be selected. This choice is essential since it will not only impact program performance, but also the time required for development as well as future maintenance costs. To avoid making a "blind" decision, all relevant evidence should be reviewed. Unfortunately, the available literature on the subject is not exhaustive.

Even though there are studies that compare translators, they do so by using sequential versions [17,21], which does not allow assessing their parallel processing capabilities. On the contrary, if they consider parallelism, they do so between languages and not between Python translators [11,13,20,22].

Based on the above, knowing the advantages and disadvantages of the different Python language translators is essential, both in sequential and multi-threaded contexts. This also applies to the primitives and functions that allow optimizing the code. Therefore, this article proposes a performance comparison between these, using the simulation of N computational bodies (*N-Body*) - a CPU-bound problem that is popular in the HPC community - as case study. This paper is an extended and thoroughly revised version of [16]. The work has been extended by providing:

– An optimized implementation in the Cython language that computes N-Body on multicore architectures, which is available in a public web repository for the benefit of the academic and industrial communities[3].
– A comparative analysis of the performance of N-Body solutions on a multicore architecture. This analysis can help Python programmers identify the strengths and weaknesses of each of them in a given situation.

The remaining sections of this article are organized as follows: in Sect. 2, the general background and other works that are related to this research are presented. Next, in Sect. 3, the implementations used are described, and in Sect. 4, the experimental work carried out is detailed and the results obtained are

---

[2] Process performed by a special class of compiler, which consists in producing source code in one language based on source code in a different language.
[3] https://github.com/Pastorsin/python-hpc-study/.

analyzed. Finally, in Sect. 5, our conclusions and possible lines of future work are presented.

2 Background

2.1 Numba

Numba is a JIT compiler that allows translating Python code into optimized machine code using LLVM[4]. According to its documentation, Numba-based algorithms can approach the speeds of those of compiled languages like C, C++, and Fortran [6], without having to rewrite its code thanks to an annotation-based approach called decorators [1].

```
1   from numba import njit
2
3   # Equivalent to indicating
4   # @jit(nopython=True)
5   @njit
6   def f(x, y):
7       return x + y
```

Fig. 1. Compilation in *nopython* mode.

JIT Compilation. The library offers two compilation modes: (1) object mode, which allows compiling code that makes use of objects; (2) *nopython* mode, which allows Numba to generate code without using the CPython API. To indicate these modes, the @jit and @njit decorators are used (see Fig. 1), respectively [6].

By default, each function is compiled at the time it is called, and it is kept in cache for future calls. However, the inclusion of the *signature* parameter will cause the function to be compiled at declaration time. In addition, this will also make it possible to indicate the types of data that the function will use and control the organization of data [6] in memory (see Fig. 2).

Multi-threading. Numba allows enabling an automatic parallelization system by setting the parameter **parallel=True**, as well as indicating an explicit parallelization through the **prange** function (see Fig. 3), which distributes the iterations between the threads in a similar way to the OpenMP **parallel for** directive. It also supports reductions, and it is responsible for identifying the variables as private to each thread if they are within the scope of the parallel zone. Unfortunately, Numba does not yet support primitives that allow controlling thread synchronization, such as semaphores or locks [6].

Vectorization. Numba delegates code auto-vectorization and SIMD instructions generation to LLVM, but it allows the programmer to control certain parameters that could affect the task at hand, such as numerical precision using the **fastmath=True** argument. It also allows using *Intel SVML* if it is available in the system [6].

---

[4] The LLVM Compiler Infrastructure, https://llvm.org/.

```
1   from numba import njit, double
2
3   @njit(double(double[::1, :],
4                double[:, ::1]))
5   def f(x, y):
6       """
7       x:  Vector 2D of the "Double" type
8           organized in columns.
9       y:  Vector 2D of the "Double" type
10          organized in rows.
11      Returns the sum of the product
12      of vectors x and y.
13      """
14      return (x * y).sum()
```

Fig. 2. Compilation in *nopython* mode with the **signature** argument

```
1    from numba import njit, double, prange
2
3    @njit(double(double[::1]), parallel=True)
4    def f(x):
5        """
6        x: 1D Vector.
7        Returns the sum of vector x
8        through a reduction.
9        """
10       N = x.shape[0]
11       z = 0
12
13       for i in prange(N):
14           z += x[i]
15
16       return z
```

Fig. 3. Compilation in *nopython* mode with the **parallel** argument

Integration with NumPy. It should be noted that Numba supports a large number of NumPy functions, which allows the programmer to control the memory organization of arrays and perform operations on them [6,7].

2.2 Cython

Cython is a static compiler for Python created with the goal of writing C code taking advantage of the simple and clean syntax of Python [2]. In other words, Cython is a Python superset that allows interacting with C functions, types, and libraries.

Compilation. As shown in Fig. 4 the Cython programming flow is very different from what the Python programmer is used to.

Fig. 4. Programming flow in Cython.

The main difference is that the file that will contain the source code has the extension .pyx unlike Python, where this extension .py. Then, this file can be compiled using a setup.py file, where compilation flags are provided to output: (1) a file with .c, extension that corresponds to the code transpiled from Cython to C, and (2) a binary file with the extension .so, that corresponds to the compilation of the C file described previously. The latter will allow importing the compiled module into any Python script.

Data Types. Cython allows declaring variables using C data types from the `cdef` statement (see Fig. 5). While this is optional, it is recommended in the documentation to optimize program execution, since it avoids the inference of CPython types at runtime. In addition, Cython allows defining the memory organization for the arrays just like Numba [2].

Multi-threading. Cython provides support for using OpenMP through the `cython.parallel`. This module contains the `prange`, function, which allows parallelizing loops using OpenMP's `parallel for` constructor. In turn, this function allows disabling GIL and defining the OpenMP *scheduling* through the `nogil` and `schedule` arguments, respectively.

It should be noted that all assignments declared within `prange` blocks are transpiled as `lastprivate`, while reductions are only identified if an *in-situ* operator is used. For example, the standard addition operation (x = x + y) will not be identified as a reduction, but the *in-situ* addition operation (x += y), will (see Fig. 6).

Vectorization. Cython delegates vectorization to the C compiler being used. Even though there are workarounds to force vectorization, it is not natively supported by Cython.

Integration with NumPy. Unfortunately, NumPy's vector operations are not supported by Cython. However, as mentioned above, NumPy can be used to control array memory organization.

```
1    cdef int x, y, z
2    cdef float a, b[100], *c
3
4    cdef struct Point:
5        double x
6        double y
```

Fig. 5. Declared variables with C data types in Cython.

```
1    from cython.parallel import prange
2
3    cdef int i
4    cdef int N = 30
5    cdef int total = 0
6
7    for i in prange(N, nogil=True):
8        total += i
```

Fig. 6. Reduction using Cython's `prange` block.

2.3   The Gravitational N-body Simulation

This problem consists in simulating the evolution of a system composed of N bodies during a time-lapse. Each body presents an initial state, given by its speed and position. The motion of the system is simulated through discrete instants of time. In each of them, every body experiences an acceleration that arises from the gravitational attraction of the rest, which affects its state.

The simulation is performed in 3 spatial dimensions and the gravitational attraction between two bodies C_i and C_j is computed according to Newtonian mechanics. Further information can be found at [18].

```
1    for t in range(STEPS):
2        for i in range(N):
3            for j in range(N):
4                Calculate the force exerted by C_j on C_i
5                Sum the forces affecting C_i
6            Calculate the displacement of  C_i
7            Move C_i
```

Fig. 7. Pseudo-code of the N-Body algorithm

The pseudo-code of the direct solution is shown in Fig. 7. This problem presents two data dependencies that can be noted in the pseudo-code. First, one body cannot move until the rest have finished calculating their interactions. Second, they cannot advance either to the next step until the others have completed the current step.

3   N-Body Implementations

In this section, the different implementations proposed are described.

3.1   CPython Implementation

Naive Implementation. Initially, a "pure" Python implementation (called *naive*) was developed following the pseudocode shown in Fig. 7, which will serve as a reference to assess the improvements introduced later. It should be noted that this implementation uses Python lists as the data structure to store the state of the bodies.

NumPy Integration. The use of NumPy arrays can speed up computation time since its core is implemented and optimized in C language. Therefore, it was decided to use these arrays as a data structure, further exploring the possible benefits of using the broadcasting function (operations between vectors provided by NumPy). Figure 8 presents the code of the CPython implementation with broadcasting.

3.2   Numba Implementation

Naive Implementation. The implementation described in Sect. 3.1, which uses the operations between vectors provided by NumPy (*broadcasting*), was selected as the initial implementation.

Numba Integration. The first Numba version was obtained by adding a decorator to the *naive* implementation (see lines 1–10 in Fig. 9a). The code was instructed to be compiled with relaxed precision using the `fastmath`, parameter, with NumPy's division model to avoid the divide-by-zero check (line 9) [6] and with *Intel SVML*, which is inferred by Numba because it is available in the system.

```
1   def nbody(N, D, positions, masses, velocities, dp):
2       # For each discrete instant of time
3       for _ in range(D):
4           # For every body that experiences a force
5           for i in range(N):
6               # Newton's Law of Universal Gravitation
7               dpos = np.subtract(positions, positions[i])
8               dsquared = (dpos ** 2.0).sum(axis=1) + SOFT
9               gm = masses * (masses[i] * GRAVITY)
10              d32 = dsquared ** -1.5
11              gm_d32 = (gm * d32).reshape(N, 1)
12              # Sum the forces
13              forces = np.multiply(gm_d32, dpos).sum(axis=0)
14
15              # Calculate acceleration of body i
16              acceleration = forces / masses[i]
17              # Calculate new velocity of body i
18              velocities[i] += acceleration * DT / 2.0
19              # Calculate new position of body i
20              dp[i] = velocities[i] * DT
21
22          # For every body that experienced a force
23          for i in range(N):
24              # Update position of body i
25              positions[i] += dp[i]
```

Fig. 8. CPython implementation with *broadcasting*.

Multi-threading. To introduce parallelism at the thread level, the `prange` statement was used. To do this, the loop that iterates over the bodies (line 5 in Fig. 8) first had to be split into two parts. The first loop is responsible for computing Newton's law of gravitational attraction and Verlet's integration, while the second one updates body position.

Arrays with Simple Data Types. NumPy's vector operations were replaced by numeric operations, and two-dimensional structures were replaced by one-dimensional ones to help Numba auto-vectorize the code (see Fig. 9).

Mathematical Operations. The following alternatives for the calculation of the denominator of Newton's universal law of attraction are proposed: (1) calculating the positive power and then dividing; and (2) multiplying by the multiplicative inverse after calculating the positive power. Additionally, the following power functions are tested: (1) `pow` function in Python's math module; and (2) `power` function provided by NumPy.

Vectorization. As noted in Sect. 2.1, Numba delegates auto-vectorization to LLVM. Even so, flags `avx512f`, `avx512dq`, `avx512cd`, `avx512bw`, `avx512vl` were defined to favor the use of this particular class of instructions.

Data Locality. To improve data locality, a version that iterates the bodies in blocks was implemented, similar to [12]. To do this, the loop on line 19 in

```
 1   @njit(
 2      void(
 3         int64,
 4         double[::1], double[::1], double[::1],
 5         double[::1],
 6         double[::1], double[::1], double[::1],
 7         double[::1], double[::1], double[::1],
 8      ),
 9      fastmath=True, parallel=True, error_model="numpy",
10   )
11   def calculate_positions(
12      N,
13      positions_x, positions_y, positions_z,
14      masses,
15      velocities_x, velocities_y, velocities_z,
16      dp_x, dp_y, dp_z,
17   ):
18      # For every body that experiences a force
19      for i in prange(N):
20         # Initialize the force of the body i
21         forces_x = forces_y = forces_z = 0.0
22         # Calculate the forces exerted on body i
23         # by every other body
24         for j in range(N):
25            # Newton's Law of Universal Gravitation
26            dpos_x = positions_x[j] - positions_x[i]
27            dpos_y = positions_y[j] - positions_y[i]
28            dpos_z = positions_z[j] - positions_z[i]
29            dsquared = (
30               (dpos_x ** 2.0) + (dpos_y ** 2.0) +
31               (dpos_z ** 2.0) + SOFT
32            )
33            gm = GRAVITY * masses[j] * masses[i]
34            d32 = dsquared ** -1.5
35            # Sum the forces
36            forces_x += gm * d32 * dpos_x
37            forces_y += gm * d32 * dpos_y
38            forces_z += gm * d32 * dpos_z
39
40         # Calculate acceleration of body i
41         aceleration_x = forces_x / masses[i]
42         aceleration_y = forces_y / masses[i]
43         aceleration_z = forces_z / masses[i]
44         # Calculate new velocity of body i
45         velocities_x[i] += aceleration_x * DT / 2.0
46         velocities_y[i] += aceleration_y * DT / 2.0
47         velocities_z[i] += aceleration_z * DT / 2.0
48         # Calculate new position of body i
49         dp_x[i] = velocities_x[i] * DT
50         dp_y[i] = velocities_y[i] * DT
51         dp_z[i] = velocities_z[i] * DT
```

(a) Function that calculates body positions.

```
 1   @njit(
 2      void(
 3         int64,
 4         double[::1], double[::1], double[::1],
 5         double[::1], double[::1], double[::1],
 6      ),
 7      fastmath=True,
 8      parallel=True,
 9      error_model="numpy",
10   )
11   def update_positions(
12      N,
13      positions_x, positions_y, positions_z,
14      dp_x, dp_y, dp_z,
15   ):
16      # For every body that experienced a force
17      for i in prange(N):
18         # Update position of body i
19         positions_x[i] += dp_x[i]
20         positions_y[i] += dp_y[i]
21         positions_z[i] += dp_z[i]
```

(b) Function that updates body positions.

Fig. 9. Numba implementation without *broadcasting*

Fig. 9a will iterate over blocks of bodies, while in two other inner loops, Newton's gravitational attraction force and Verlet's integration will be calculated, respectively.

Threading Layer. Thread API changes were made through the *threading layers* that Numba uses to translate parallel regions. To do this, different options were tested: *default, workqueue, omp* (OpenMP) and **threading**. For the first three, the source code did not have to be modified, since Numba is responsible for translating the **prange** block to the selected API. However, this was not the case with **threading** - thread distribution had to be coded together with the specification of the parameter **nogil=True** to disable GIL. It should be noted that the *tbb* option was not used because it was not available in the support server.

3.3   Cython Implementation

Naive Implementation. As initial implementation (*naive*), the one described in Sect. 3.1 was used. This implementation uses various NumPy arrays as data structures, which allows more flexible management of memory, particularly concerning organization and data types.

Cython Integration. No changes were made to the code of the *naive* implementation (see Sect. 3.3), but it was compiled using Cython. To do so, the extension of the code was simply changed from `.py` to `.pyx`.

Explicit Typing. As shown in Fig. 10 the data types provided by Cython were explicitly defined to reduce interaction with the CPython API. Initially, to reduce unnecessary checks at runtime, the following compiler directives are provided (lines 1–4) [2]:

- *boundcheck* (line 1) avoids index error verifications on arrays.
- *wraparound* (line 2) prevents arrays from being indexed relative to the end. For example, in Python, if A is an array with statement `A[-1]`, its last element can be obtained.
- *nonecheck* (line 3) avoids verifications due to variables that can potentially take the value *None*.
- *cdivision* (line 4) performs the division through C avoiding CPython's API. CPython.

On line 5, a hybrid function type is indicated through the `cpdef` statement, which allows the function to be imported from other applications developed in Python. Then, on lines 6–10, the Cython data types that will later be transpiled to C data types are specified. In particular, arrays are specified with the `double[::1]`, data type, which ensures that the arguments received are NumPy arrays contiguous in memory [2]. Finally, on lines 12–16, the data types corresponding to local variables are declared using the `cdef` statement.

Multi-threading. This version introduces thread-level parallelism through the `prange` statement provided by Cython. To do this, `range` statements were replaced by `prange` statements. In particular, the instruction was to use the *static* policy as *schedule* to evenly distribute the workload among the threads considering computation regularity. Additionally, GIL was disabled through the *nogil* argument to allow these to be executed in parallel (see Fig. 10).

Finally, it should be noted that the `prange` statement is transpiled into an OpenMP `parallel for` [2]. Therefore, it has an implicit barrier that allows threads to be synchronized to comply with the data dependencies described in Fig. 7.

Mathematical Operations. A decision was made to evaluate the same alternatives for calculating the denominator for Newton's universal law of attraction as those described in Sect. 3.2. In particular, no changes were made to the power functions to avoid interaction with CPython's API.

Vectorization. As mentioned in Sect. 2.2, Cython delegates auto-vectorization to the C compiler. However, the flags `-xCORE-AVX512`, `-qopt-zmm-usage=high`, `-march=native` were provided to favor the use of AVX-512 instructions.

Data Locality. To better take advantage of cache, a variant that iterates the bodies in blocks was implemented, similar to the one described in Sect. 3.2. In this version, the loop on line 21 in Fig. 10 is split into two other inner loops. The first one calculates Newton's gravitational attraction force, while the second one calculates displacement using the *velocity verlet* integration method.

4   Experimental Results

4.1   Experimental Design

All tests were carried out on a Dell Poweredge server equipped with $2\times$ Intel Xeon Platinum 8276's with 28 cores (2 hw threads per core) and 256 GB of RAM. The operating system was Ubuntu 20.04.2 LTS, and the translators and libraries used were Python v3.8.10, PyPy v7.3.1, NumPy v1.20.1, Numba v0.52.0, Cython v0.29.22 and ICC v19.1.0.166.

For implementation evaluation, different workloads ($N = \{256, 512, 1024, 2048, 4096, 8192, 16384, 32768, 65536, 131072, 262144, 524288\}$) and number of threads ($T = \{1, 56, 112\}$) were used. The number of simulation steps remained fixed ($I=100$). Each proposed optimization was applied and evaluated incrementally based on the initial version[5]. To evaluate performance, the GFLOPS (billion FLOPS) metric is used, with equation $GFLOPS = \frac{20 \times N^2 \times I}{t \times 10^9}$, where N is the number of bodies, I is the number of steps, t is execution time (in seconds) and factor 20 represents the number of floating-point operations required by each interaction[6].

4.2   CPython Performance

Figure 11 shows the performance obtained with the *naive* version with NumPy when varying N. As it can be seen, the incorporation of NumPy arrays without using broadcasting worsened the performance by $2.9\times$ on average. Because values are stored directly in NumPy arrays and must be converted to Python objects when accessed, unnecessary conversions are carried out using this algorithm. In

---

[5] Each previous version is labeled as *Reference* in all graphics.
[6] A widely accepted convention in the literature for this problem.

```
1    @boundscheck(False)
2    @wraparound(False)
3    @nonecheck(False)
4    @cdivision(True)
5    cpdef void nbody(
6        int N, int D, int T,
7        double[::1] positions_x, double[::1] positions_y, double[::1] positions_z,
8        double[::1] masses,
9        double[::1] velocities_x, double[::1] velocities_y, double[::1] velocities_z,
10       double[::1] dp_x, double[::1] dp_y, double[::1] dp_z,
11   ):
12       cdef double forces_x, forces_y, forces_z
13       cdef double aceleration_x, aceleration_y, aceleration_z
14       cdef double dpos_x, dpos_y, dpos_z
15       cdef double dsquared, gm, d32
16       cdef int i, j
17
18       # For each discrete instant of time
19       for _ in range(D):
20           # For every body that experiences a force
21           for i in prange(N, nogil=True, schedule="static", num_threads=T):
22               # Initialize the force of the body i
23               forces_x = forces_y = forces_z = 0.0
24
25               # Calculate the forces exerted on body i by every other body
26               for j in range(N):
27                   # Newton's Law of Universal Gravitation
28                   dpos_x = positions_x[j] - positions_x[i]
29                   dpos_y = positions_y[j] - positions_y[i]
30                   dpos_z = positions_z[j] - positions_z[i]
31                   dsquared = (dpos_x ** 2.0) + (dpos_y ** 2.0)
32                       + (dpos_z ** 2.0) + SOFT
33                   gm = GRAVITY * masses[j] * masses[i]
34                   d32 = dsquared ** -1.5
35                   # Sum the forces
36                   forces_x = forces_x + gm * d32 * dpos_x
37                   forces_y = forces_y + gm * d32 * dpos_y
38                   forces_z = forces_z + gm * d32 * dpos_z
39
40               # Calculate acceleration of body i
41               aceleration_x = forces_x / masses[i]
42               aceleration_y = forces_y / masses[i]
43               aceleration_z = forces_z / masses[i]
44               # Calculate new velocity of body i
45               velocities_x[i] += aceleration_x * DT / 2.0
46               velocities_y[i] += aceleration_y * DT / 2.0
47               velocities_z[i] += aceleration_z * DT / 2.0
48               # Calculate new position of body i
49               dp_x[i] = velocities_x[i] * DT
50               dp_y[i] = velocities_y[i] * DT
51               dp_z[i] = velocities_z[i] * DT
52
53           # For every body that experienced a force
54           for i in prange(N, nogil=True, schedule="static", num_threads=T):
55               # Update position of body i
56               positions_x[i] += dp_x[i]
57               positions_y[i] += dp_y[i]
58               positions_z[i] += dp_z[i]
```

Fig. 10. Parallel Cython implementation.

contrast, this does not happen in the *naive* version because values are already saved directly as Python objects.

This issue was solved by adding *broadcasting*; that is, by performing vector operations between NumPy arrays. This avoids unnecessary conversions because operations are carried out internally in the NumPy core [7]. As it can be seen, performance improved by 10× on average with respect to the *naive* version.

Finally, Fig. 12 shows the performance of CPython compared to PyPy. As it can be seen, CPython's performance tends to improve as size increases, whereas PyPy's[7] performance remains constant.

---

[7] The implementation executed with PyPy was the naive version since the other versions use NumPy arrays and PyPy is unable to optimize them.

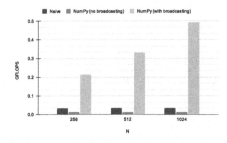

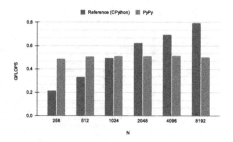

Fig. 11. CPython - performance obtained with the different versions for various values of N.

Fig. 12. PyPy - performance obtained for various values of N.

4.3 Numba Performance

Figure 13 shows the performance when activating the compilation options and applying multi-threading with various values of N. Even though the Numba compilation options (`njit+fastmath+svml`) have practically no effect on the performance of this version, there is a significant improvement when using threads to compute the problem. In particular, an average improvement of 33× and 38× for 56 and 112 threads, respectively, can be noted.

Figure 14 shows the significant improvement obtained when using arrays with simple data types instead of compound ones (an average of 41× in the case of 112 threads). Even though the second one simplifies coding, it also involves organizing the data in the form of an *array of structures*, which imposes limitations on the use of the SIMD capabilities of the processor [14]. Additionally, it can also be noted that using *hyper-threading* results in an improvement of approximately 78% in this case.

There are practically no changes in performance if the mathematical calculations and power functions used are different from those described in Sect. 3.2 (see Fig. 15). This is because no matter which option is used, the resulting machine code is always the same. Something similar happens when explicitly specifying the use of AVX-512 instructions. As mentioned in Sect. 2.1, Numba attempts to auto-vectorize the code via LLVM. Looking at the machine code, it was observed that the generated instructions already made use of these extensions.

The processing by blocks described in Sect. 3.2 did not improve the performance of the solution, as it can be seen in Fig. 16. The performance loss is related to the fact that this computation reorganization produces failures in LLVM when auto-vectorizing. Unfortunately, since Numba does not offer primitives to specify the use of SIMD instructions explicitly, there is no way to fix this.

Figure 17 shows the performance obtained for precision reduction with various data types and workloads (N). It can be seen that the use of the `float32` data type (instead of `float64`) leads to an improvement of up to 2.8× GFLOPS, at the cost of a reduction in precision. Similarly, relaxed precision produced a significant acceleration on both data types: `float32` (17.2× on average) and `float64` (11.4× on average). In particular, the performance peak is 1524/536

GFLOPS in single/double precision. It is important to mention that this version achieves an acceleration of 687× compared to the *naive* implementation (`float64`).

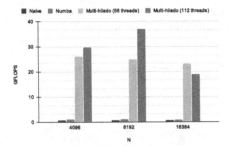

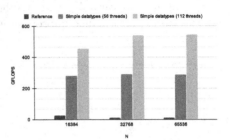

Fig. 13. Numba - performance obtained for compiling and multi-threading options for various values of N.

Fig. 14. Numba - performance obtained with the different versions for various values of N.

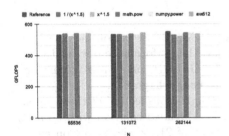

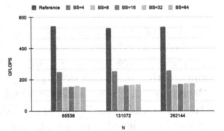

Fig. 15. Numba - performance obtained using different mathematical calculations, power functions and AVX512 instructions for various values of N.

Fig. 16. Numba - performance obtained when processing in blocks for various values of N.

Finally, Fig. 18 shows that, when using 112 threads, OpenMP and the default *threading layer* provided by Numba outperform the others by an average of 9.3%. In turn, it can be seen that the use of `threading` is below all tested threading layers. This is due to the overhead generated by the use of Python objects to synchronize the threads, which does not occur in the other threading layers because synchronization is carried out on its own API.

4.4   Cython Performance

Figure 19 shows that the Cython integration (without specifying the data types for the variables) did not produce a significant improvement concerning the *naive* version, since the resulting code is using CPython's API. On the other hand, defining the variables with Cython data types reduces this interaction and, therefore, performance improves remarkably (547.7× on average).

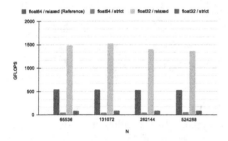

Fig. 17. Numba - performance obtained for precision reduction for various data types and values of N.

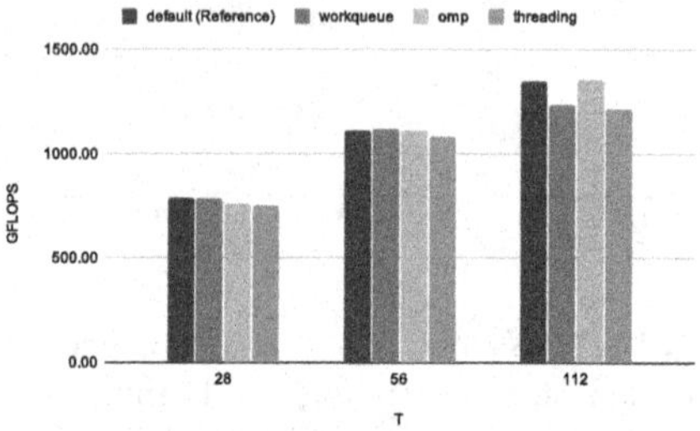

Fig. 18. Numba - performance obtained with the different *threading layers* in Numba for various values of T and a fixed value of $N = 524288$.

As shown in Fig. 20, specifying the use of AVX-512 instructions achieved an improvement of 1.7× on average. In particular, the performance obtained with the `-march=native` flag was slightly higher (1.4 GFLOPS on average) than that obtained with the `-xCORE-AVX512 -qopt-zmm-usage=high` flag.

Both multi-threading and mathematical optimizations led to positive results. On the one hand, Fig. 21 shows that the multi-threaded solution with 56 and 112 threads achieved a remarkable improvement of 21.1× and 34.6× on average, respectively. On the other hand, Fig. 22 shows the performance obtained when applying the mathematical operations described in Sect. 3.3. As it can be seen, using a direct division degraded the performance by 41%; while calculating the multiplicative inverse by positive power did not result in any significant improvements.

Block processing significantly worsened solution performance for all tested block sizes (see Fig. 23). This is because the compiler identifies false dependencies in the code and does not generate the corresponding SIMD instructions. Unfortunately, this cannot be fixed, as Cython does not provide a way to tell the compiler that it is safe to vectorize operations.

Finally, Fig. 24 shows that precision reduction had practically no effect on the performance obtained. However, using *float* as data type improved performance noticeably (1362 GFLOPS on average) at the cost of less representation in the final result.

4.5   Performance Comparison

First of all, it should be noted that the versions with CPython and PyPy were not included in the final comparison due to their low performance (0.5 GFLOPS on average). Figure 25 shows a comparison between the optimized implementations of Numba and Cython for various workloads and data types. As it can be seen, when using double precision, Cython was slightly faster than Numba by an average of 16.7 GFLOPS; while in single precision, Numba was superior by an average of 73 GFLOPS. These values represent improvements of 3% and 5%, respectively. In turn, it is important to mention that both final versions of Numba and Cython achieved an average acceleration of 1018× and 1050× respectively, compared to the best CPython implementation (`float64`).

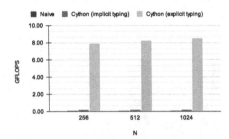

Fig. 19. Cython - performance obtained with and without explicit typing for various values of N.

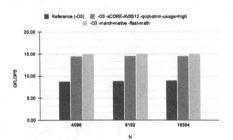

Fig. 20. Cython - performance obtained with the different compilation options for various values of N.

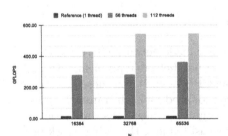

Fig. 21. Cython - performance obtained with the multi-threaded solution for various values of N.

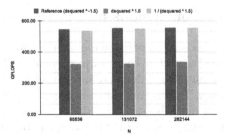

Fig. 22. Cython - performance obtained with the mathematical calculations for various values of N.

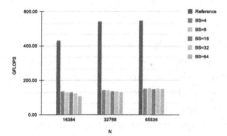
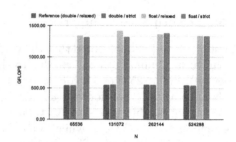

Fig. 23. Cython - performance obtained when processing in blocks for various values of N.

Fig. 24. Cython - Performance obtained for precision reduction for various data types and values of N.

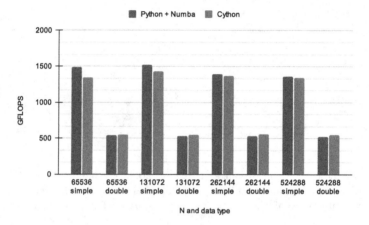

Fig. 25. Performance comparison between the final versions of Numba and Cython for various data types and values of N.

5   Conclusions and Future Work

In this work, a performance comparison has been made between CPython, PyPy, Numba, and Cython. In particular, N-Body -a parallelizable problem with high computational demand and considered to be CPU-bound- was chosen as a case study. To this end, different algorithms were produced for each translator, starting from a base version and applying incremental optimizations until reaching the final version. In this sense, the benefits of using multi-threading, block processing, *broadcasting*, different mathematical calculations and power functions, vectorization, explicit typing, and different thread APIs were explored.

Considering the results obtained, it can be said that there were no significant differences between the performance of Numba and Cython. However, both translators significantly improved CPython's performance. This was not the case with PyPy, since it failed to improve the performance of CPython+NumPy due to its inability to parallelize, associated with GIL. Therefore, it can be stated that in contexts similar to those of this study, both Numba and Cython can be powerful tools to accelerate CPU-bound applications developed in Python. The choice between one or the other will be largely determined by the approach that the development team finds most convenient, considering the characteristics of each one.

As future work, it would be interesting to extend on the following directions:

– Replicating the study carried out considering: (1) other case studies that are computationally intensive but whose characteristics are different from those of *N-Body*; (2) other multicore architectures different from the one used in this work.
– Considering that programming effort is an increasingly relevant issue [5], comparing the solutions developed from this perspective.
– Given that other technologies allow parallelism to be implemented at the process level in Python, comparing these considering not only their performance but also programming cost.

References

1. Decorators - Python Tips 0.1 documentation. https://book.pythontips.com/en/latest/decorators.html
2. Cython: C-Extensions for Python. https://cython.org/
3. Home - OpenMP. https://www.openmp.org/
4. Home — Jython. https://www.jython.org/
5. Microsoft's new research lab studies developer productivity and well-being — VentureBeat. https://venturebeat.com/2021/05/25/microsofts-new-research-lab-studies-developer-productivity-and-well-being/
6. Numba documentation - Numba 0.53.1-py3.7-linux-x86_64.egg documentation. https://numba.readthedocs.io/en/stable/index.html
7. NumPy. https://numpy.org/
8. PyPy. https://www.pypy.org/
9. Sunsetting Python 2 — Python.org. https://www.python.org/doc/sunset-python-2/
10. What Is the Python Global Interpreter Lock (GIL)? - Real Python. https://realpython.com/python-gil/
11. Cai, X., Langtangen, H.P., Moe, H.: On the performance of the python programming language for serial and parallel scientific computations. Sci. Program. **13**(1), 31–56 (2005). https://doi.org/10.1155/2005/619804
12. Costanzo, M., Rucci, E., Naiouf, M., Giusti, A.D.: Performance vs programming effort between rust and C on multicore architectures: case study in N-body. In: 2021 XLVII Latin American Computer Conference (CLEI) (2021). https://doi.org/10.1109/CLEI53233.2021.9640225
13. Gmys, J., Carneiro, T., Melab, N., Talbi, E.G., Tuyttens, D.: A comparative study of high-productivity high-performance programming languages for parallel metaheuristics. Swarm Evol. Comput. **57**, 100720 (2020). https://doi.org/10.1016/j.swevo.2020.100720
14. Intel Corp.: How to manipulate data structure to optimize memory use on 32-bit intel® architecture (2018). https://tinyurl.com/26h62f76
15. Marowka, A.: Python accelerators for high-performance computing. J. Supercomput. **74**(4), 1449–1460 (2017). https://doi.org/10.1007/s11227-017-2213-5

16. Milla, A., Rucci, E.: Acelerando Código Científico en Python usando Numba. XXVII Congreso Argentino de Ciencias de la Computación (CACIC 2021), p. 12, October 2021. http://sedici.unlp.edu.ar/handle/10915/126012

17. Roghult, A.: Benchmarking Python interpreters: measuring performance of CPython, Cython, Jython and PyPy. Master's thesis, School of Computer Science and Communication, Royal Institute of Technology, Sweden (2016)

18. Rucci, E., Moreno, E., Pousa, A., Chichizola, F.: Optimization of the N-body simulation on intel's architectures based on AVX-512 instruction set. In: Pesado, P., Arroyo, M. (eds.) CACIC 2019. CCIS, vol. 1184, pp. 37–52. Springer, Cham (2020). https://doi.org/10.1007/978-3-030-48325-8_3

19. TIOBE Software BV: TIOBE Index for November 2021 (2021). https://www.tiobe.com/tiobe-index/

20. Varsha, M., Yashashree, S., Ramdas, D.K., Alex, S.A.: A review of existing approaches to increase the computational speed of the Python language. Int. J. Res. Eng. Sci. Manag. (2019)

21. Wilbers, I., Langtangen, H.P., Odegard, A.: Using cython to speed up numerical python programs. In: Proceedings of MekIT, pp. 495–512 (2009)

22. Wilkens, F.: Evaluation of performance and productivity metrics of potential programming languages in the HPC environment. Bachelor's thesis, Faculty of Mathematics, Informatics und Natural Sciences, University of Hamburg, Germany (2015)

Computer Technology Applied to Education

Administrative Modular Integral System Participatory. Design in Software Development

Verónica K. Pagnoni[1]([✉]) [iD], Diego F. Craig[2] [iD], Juan P. Méndez[3] [iD],
and Eduardo E. Mendoza[3] [iD]

[1] Higher Institute of Teacher Training 'Bella Vista', Corrientes, Argentina
equipotecnico@dgescorrientes.net
[2] Ministry of Education of the Corrientes Province - Higher Education Administration,
Buenos Aires, Argentina
[3] Ministry of Education of the Corrientes Province – Systems Management,
Buenos Aires, Argentina

Abstract. The Modular Integral Administrative System (SIMA – Spanish initials) is a computer application that has functionalities for the integral management of administrative data of the ISFDs(ISFD – Superior Institute for Teacher Training. The Argentinian government creates and funds these institutions to educate and graduate professionals for certain careers, mainly teachers for necessary areas.) of the Province of Corrientes, developed as a priority line of the Technology of the Information and Communication (TIC) of DNS(DNS - Higher Level Education Directorate – The Jurisdiction's governmental body in charge of directing and regulating the Higher-Level Educational Sector of the prov-ince (The province of Corrientes, in this case).) of the provincial Ministry of Education, in the Argentine Republic. In this paper, the process, elaborated and implemented by the SIMA development team, in accordance with the Participatory Design paradigm, is exposed. Future lines of work are also expressed.

Keywords: Software development · Participatory design · Educational management

1 Introduction

This article starts from the characterization of SIMA as an SIGED (Educational Information and Management System – Spanish Initials) exposed in [1] and then exposes the process that is carried out for its development using within the Participatory Design paradigm.

Computer systems for data management and the execution of administrative tasks have been widely used alternatives in public administration in recent decades. Computerization makes it possible to streamline and make more efficient the processes that administrative work entails. [2–4].

SIMA is a computer system that was developed within the Bella ISFD - in the city of Bella Vista, Province of Corrientes, as a computer solution for the management of

P. Pesado and G. Gil (Eds.): CACIC 2021, CCIS 1584, pp. 41–48, 2022.
https://doi.org/10.1007/978-3-031-05903-2_3

the administrative tasks of this institution; since 2018 it begins to be implemented at the jurisdictional level, offering services to various institutes in the province.

This system is based on four pillars: modular programming, constant adaptation to current regulations, user participation and adaptation to changes. It has a series of functionalities that cover a wide range of activities, aimed at providing support to different people involved in Higher Level Education. It is implemented in a technological infrastructure arranged in the facilities of the Technological Systems Directorate, the DNS and in the different ISFDs that use it, all part of the Ministry of Education of the Province of Corrientes, Republic of Argentina [1].

2   Context

As mentioned in the introduction, the SIMA arises as an initiative of a particular unit to later form a transcendental line of work of the TIC area of the DNS of the Province of Corrientes. Thus since 2018, its development continues within the Computerization Project of ISFD of the Province of Corrientes.

The TIC area of the DNS aims to "expedite and strengthen the management of information that circulates inside and outside higher education institutes" [1]. As SIMA, it aims to be a solution tailored to the needs of Higher Educational Level institutions, facilitating the performance of administrative tasks, and favoring the management and protection of data.

3   Theoretical Foundations on Which SIMA is Developed

It is agreed [5] that the use of technologies allows the possibility of systematizing processes that are carried out manually or semi-automatically. Therefore, digital transformation can promote innovation in educational management.

As expressed in [6], the incorporation of technologies in public administration "seeks ways to find mechanisms that have a positive impact on daily work and the generation of modern responses to the demands of society".

In [7] an Educational Information and Management System (SIGED) is defined as "as the set of key processes of educational management that serve to, register, exploit, generate and disseminate strategic information online in an integral way, framed by a concrete legal, institutional and technological infrastructure".

As mentioned in [8], the advantages offered by a SIGED are:

1. The availability of timely and quality information for the design of policies and the allocation of resources.
2. The time savings resulting from those administrative tasks that go from being carried out manually to being implemented using technologies.
3. Budget savings due to more efficient use of resources.

Other benefits can be added such as:

4. The possibility of accessing useful information by any user of the system, without space–time limitations, using technologies.
5. Improvements in equal opportunities by publicizing educational proposals and facilitating the processes of entry, access and permanence in the higher education subsystem, both for teachers and students.

The software development industry, like so many others, has evolved rapidly in recent years with the emergence of the Internet and tools that make it possible to build systems collaboratively and quickly. In this context, traditional software development methodologies have been adapted and new ones have emerged [9].

According to [10] some software implementations fail due to "lack of information from users, lack of adequate training, dissatisfaction due to issues associated with the daily task they perform, lack of effective communication within the organization". That is why the implementation of participative design applied to software development is considered an important strategy for the success of a system. It agrees with [11] that participatory design in software development processes is a method that proposes the optimal use of the knowledge possessed by users, achieving their commitment and satisfaction by being part of the product construction process. As stated in [1] "user participation in all stages of system development has become crucial, in the identification of requirements, in the definition of the objects that belong to the problem domain and in the efficient implementation of the problem system".

What was established in [12] was considered, where the authors propose the division of a complex software system into simple parts or modules. Modular programming makes it possible for multiple developers to divide up the work and independently program or debug different portions of the system. In addition, the modules force to maintain logical links between the various components and optimize the maintenance work. Besides, it is important to mention that programming by modules allows a design that minimizes the dependencies between the parts of the software [13–15].

As stated in [1], "it is considered important to highlight that in these times of vertiginous changes it is essential to carry out software development that adapts to them, establishing ways of working that favor the optimization of processes". In this sense, modular programming and the active participation of the user in the construction of the software product help to achieve dynamic development.

4 SIMA Platform Architecture

As mentioned in [1] SIMA is based on four pillars:

- Constant adaptation to current regulations: The DNS brings together and governs the actions of the Higher Institutes of Teacher and Technical Training of the entire Province of Corrientes. The DNS issues regulations as problems, needs or new guidelines arise at the jurisdictional or national level. As stated in [1] "its study and analysis are essential to establish the functionalities in SIMA that will help to its fulfillment".

- Adaptation to changes: The modifications in SIMA are not only given by the need to comply with the changing regulations; there are other factors that cause modifications in the programming of the modules. An important influence is given by achieving a system that helps the user to carry out their work in the most optimal way. For this reason, the functionalities are adjusted until the user expresses their agreement and tools become adapted to their needs for the correct development of their work. It is also a permanent concern to make SIMA an accessible software, which implies that certain modifications are made so that it can be operated by as many users as possible regardless of their abilities.
- Modular programming: Due to the continuous variations, SIMA must have a program that can be easily modified and allow for constant adaptation. Besides, given that SIMA is a platform that is permanently used by hundreds of users, it cannot be out of service for a long time. Therefore, it a modular programming was chosen, since it allows the addition of functionalities, and likewise, makes it possible to change a specific functionality without modifying the others that are in use.
- User Participation: To arrive at the definition and adjustment of the SIMA functionalities, hard work is carried out with the users, with whom the needs to be satisfied are defined, analyzed and programmed, and then their opinion is heard again to make the necessary modifications.

As expected, the computerization needs are many, the priority in its programming is given mainly by the following factors:

1. Programming of functionalities to fulfill the tasks established in the school calendar.
2. Emergence of a new regulation.
3. Needs expressed by several referents or other users in general meetings.
4. Improvements or adjustments to functions already programmed.

5   Application of Participatory Design

As stated in [1] "user participation is key both in the implementation of SIMA and in its development", they are constantly consulted to "gather information on the priorities to be computerized, the shortcomings that must be corrected and the improvements that can be implemented".

In the development of SIMA, different types of users were considered:

- Superior Level: this body participates by issuing regulations that must be implemented in the system.
- Directors: Principals.
- Referents: teachers of an institution that fulfill the role of communicators of queries, doubts, or needs when general meetings are not held.
- Final users who fulfill a certain function: secretaries, administrative staff, guards, teachers, students.

In Fig. 1 you can see the different users and their interaction with SIMA:

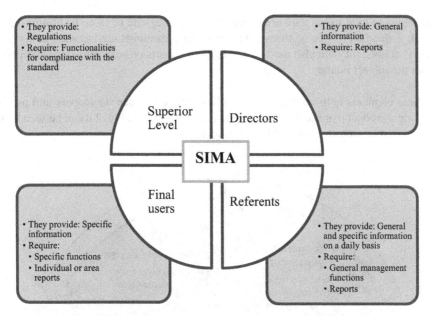

Fig. 1. Users and their interaction with SIMA

From its beginning to the present, the team in charge of development of SIMA has been implementing different strategies to achieve the definitions and adjustments of the functionalities of the system in which user participation has been fundamental. These are now constituted in a process, framed in the Participatory Design paradigm.

Participatory design is a user-centered method, in which participants are invited to collaborate throughout the software development process [15]:

- In the initial definition of requirements.
- Collaborating to establish possible solutions.
- Providing feedback after the use of the functionalities.

The tools used to achieve user participation at all stages of SIMA development are:

- General meetings to define requirements: these are held in person or by videoconference. Users of different profiles are invited.
- Special meetings with users: meetings held with users who perform a specific function to gather information on how specific tasks are carried out.
- Communication with referents: this occurs practically daily.
- Advance meetings: meetings with users where possible functionalities are shown to adjust their programming.
- "Feedback" meetings; they invite users who may or may not be in contact with the development team. In these meetings, the developers only participate as listeners, the other participants talk about a topic and there is a moderator who guides the exchange. Participants should focus on defining the pros and cons of a certain functionality.

– Meetings to train users: they are organized by inviting one or more institutions and work is done on a specific theme. Although these meetings are intended to correct users' doubts, they are also used to define requirements or improvements associated with the subject matter.

These elements help interaction and collaboration between developers and users, achieving a product that is constantly updated and evolving. In Fig. 2 it can be seen how they are used in each stage of software development:

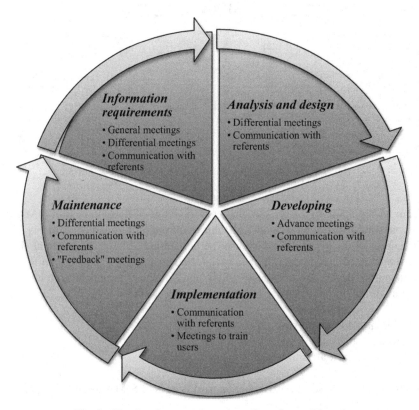

Fig. 2. User involvement throughout the software lifecycle

In this way, SIMA is built for and with users, who provide their knowledge and collaboration in all stages of the application life cycle. In this way, a collaborative process is achieved where all the participants acquire relevance and develop a sense of belonging to the project.

6 Conclusions and Future Lines

SIMA is a computer system that has arisen in response to the specific needs of an ISFD, to soon become a tool used by the majority of ISFDs in the Province of Corrientes.

The Modular Integral System of Educational Administration has become a R&D project through which a process of survey of needs, analysis, design, and development of a platform for the integration of TIC oriented to the administrative management of educational institutions of Superior Level of the Province has been accomplished.

SIMA is consolidated as a SIGED, considering its architecture and the multiplicity of functionalities it offers. As expressed in [1], there is still a need to improve the technological infrastructure of the headquarters and the institutions to achieve better connectivity, the adjustment of user interfaces and the tools proposed to achieve web accessibility for all users.

SIMA is built with and for users using techniques to promote Participatory Design. This article has exposed the development process used and its elements.

In the future, it is expected to adjust, complete and improve the described process, implementing new communication and collaboration tools, to enhance the development of this software.

References

1. Mendoza, E.F., Méndez, J.P., Craig, D.F., Pagnoni, V.K.: SIMA. Un sistema integral modular para la gestión administrativa de la Educación Superior. Congreso Argentino de Ciencias de la Computación (2021)
2. Farabollini, G.R.: Gobierno electrónico: una oportunidad para el cambio en la administración pública. Segundo Congreso Argentino de Administración Pública. Sociedad, Estado y Administración. Asociación Argentina de Estudios de Administración Pública Asociación de Administradores Gubernamentales (2003)
3. Rodríguez Conde, M.J.: Aplicación de las TIC a la evaluación de alumnos universitarios. Revista Teoría de la educación: educación y cultura en la sociedad de la información. Salamanca vol. 6, no. 2 (2005)
4. Álvarez Mayor, J.A., Díaz Marín, A.: Las TIC como componente dinamizador para la gestión del conocimiento de las empresas en perfeccionamiento empresarial en Cuba. Revista de Investigación Latinoamericana en Competitividad Organizacional, ISSN-e 2659–5494, N°. 11, 2021, pp. 65–76 (2021).
5. Pombo, C., Gupta, R., Stankovic, M.: Servicios sociales para ciudadanos digitales: Oportunidades para América Latina y el Caribe. BID, Washington (2018)
6. Lacunza, A.C., Clark, R.I., Marafuschi Phillips, M.A.: La Gestión documental electrónica en la UNLP. El camino hacia el expediente electrónico. Revista ES (en y sobre Educación Superior) 1(1–2), e022 (2021)
7. Arias Ortiz, E., Eusebio, J., Pérez Alfaro, M., Vásquez, M., Zoido, P.: Del papel a la nube: Cómo guiar la transformación digital de los Sistemas de Información y Gestión Educativa (SIGED). BID, Washington (2019)
8. Arias Ortiz, E., Eusebio, J., Pérez Alfaro, M., Vásquez, M., Zoido, P.: Los Sistemas de Información y Gestión Educativa (SIGED) de América Latina y el Caribe: la ruta hacia la transformación digital de la gestión educativa. BID, Washington (2021)
9. Maida, E.G., Pacienzia, J.: Metodologías de desarrollo de software. Tesis de Licenciatura en Sistemas y Computación. Facultad de Química e Ingeniería "Fray Rogelio Bacon". Universidad Católica Argentina (2015)
10. Panizzi, M.: Propuesta de Recomendaciones para la Implementación de Sistemas Informáticos Basadas en el Enfoque Socio-técnico y el Diseño Participativo. Revista Latinoamericana de Ingeniería de Software 3(1), 1–40 (2015). ISSN 2314–264

11. Estayno, M.G., Panizzi, M.D.: Participación de los usuarios en el proceso de desarrollo de software. Workshop de Investigadores en Ciencias de la Computación (2009)
12. Ghezzi, C., Jazayeri, M., Mandrioli, D.D.: Fundamentals of Software Engineering. Prentice-Hall, Hoboken (2002)
13. Cháves, T.: Programación Modular. Universidad Autónoma del Estado de México (2017)
14. EKCIT Software modular. https://www.ticportal.es/glosario-tic/software-modular. Accessed 02 Feb 2022
15. Feller, L., Hoare, M.E., Portiere, L. Diseñando para los usuarios o diseñando con los usuarios?: el diseño participativo como paradigma de un proyecto. Interaction South America (ISA 14): 6ta. Conferencia Lationamericana de Diseño de Interacción (2014)

Systematic Review of Educational Methodologies Implemented During the COVID-19 Pandemic in Higher Education in Ibero-America

Omar Spandre[1]([✉]), Paula Dieser[1], and Cecilia Sanz[2,3]

[1] Application of Computer Technology in Education. School of Computer Sciences, UNLP, La Plata, Argentine
spandreomar@gmail.com

[2] Institute of Research in Computer Sciences LIDI – CIC. School of Computer Sciences, UNLP, La Plata, Argentine
csanz@lidi.info.unlp.edu.ar

[3] Scientific Research Commission from the Province of Buenos Aires, Buenos Aires, Argentine

Abstract. Since the interruption of face-to-face classes due to the COVID-19 pandemic, Higher Education institutions adapted their courses to a virtual format in order to serve students during the contingency. This work analyzes, through a systematic review of a twenty-four-article corpus, which study the pedagogical models adopted by such Institutions in Ibero-America. The reviewed works have been published between 2020 and 2021, and analyze the implications of reorganizing the teaching-learning processes, based on the framework of the Community of Inquiry model. The results show that a high percentage of the research focuses on Teaching presence, especially in the educational and organizational design, and much less focuses on Cognitive presence, and a limited production on the Social presence dimension and the analysis of indicators that reinforce learning.

Keywords: Remote learning · Virtual education · COVID-19 · Higher education · Community of inquiry model

1 Introduction

This work analyzes the emerging crisis, especially in Higher Education, as a consequence of the suspension of face-to-face classes due to the COVID-19 pandemic, through a systematic review (SR) of a twenty-four-article corpus published between 2020 and 2021 in Ibero-America [1–24].

The new coronavirus SARS-CoV-2 and its potential illness COVID-19 was determined as a pandemic of levels with no precedents, which nowadays, more exactly, has entered a new advanced and more lethal phase [25].

The challenge for the educational systems during the last months has been to sustain education vitality and to promote the development of meaningful learning. For this

P. Pesado and G. Gil (Eds.): CACIC 2021, CCIS 1584, pp. 49–63, 2022.
https://doi.org/10.1007/978-3-031-05903-2_4

purpose, it has relied on two key allies: its teachers and virtuality; more precisely, teachers through virtuality [7].

Estimates from the UNESCO International Institute for Higher Education in Latin America and the Caribbean (IESALC) show that the temporary closure of Higher Education Institutions (HEIs) had affected approximately 23,4 million higher education students and 1,4 million teachers in Latin America and the Caribbean even before March 2020. This represented approximately more than 98% of the region's population of higher education students and teachers in the region [26].

In this health emergency, it is important to keep each person´s health safe but, even so, it is also important to continue with education. For this purpose, HEIs have adopted certain tools and platforms [10].

Most of the teachers were not prepared for remote learning but, with courage and determination, they made an effort to understand the meaning of remote teaching using a fully online learning environment, struggling to create content that was engaging and relevant, or experimenting with digital assessment. As teachers chose to play safe and avoided major risks, most of them simply limited their activity to replicate their traditional classroom experiences, holding online conferences through web conference systems such as Zoom, Skype, Microsoft Teams, Google Meet and WhatsApp, and assessment practices based on online exams. This excessive simplification of remote and online teaching methodologies has resulted in an excessive content delivery-based approach, which depreciates adequate support and feedback that are important to ensure students 'performance [14].

According to experts, most of these practices are better characterized as an emergency remote teaching, defined as "temporary shift of instructional delivery to an alternate delivery mode due to crisis circumstances" [27].

Some devices and tools implemented to deal with the crisis account for digital technologies scope and usefulness, and set trends among the main decisions made in the HEIs such as the virtual reality, video game-based learning, e-learning, artificial intelligence, online and mobile education, as well as resources considered supports to their viability, for example digital printers, virtual teaching and learning environment (VTLE) and interactive digital boards, the aims of which is the dynamic and transformative modelling in the management of the HEIs [17].

All this arsenal of innovative tools adapted to deal with the crisis, or the didactic-pedagogical strategies implemented by HEIs on many situations do not have into account neither the immediate effects nor the impact that will be produced by the transition to emergency remote education. Other no less important impacts on the different actors, less visible and not yet documented, include areas such as socio-emotional, labor, financial and, obviously, about the functioning of the system as a whole [18].

The before-mentioned aspects motivated this work, in which a systematic review (SR) is achieved in order to analyze the selection of articles within the framework of the Community of Inquiry Model (CoI) by Garrison and Anderson [28]. Thus, based on its dimensions: Teaching Presence, Cognitive Presence and Social Presence, the pedagogical-didactic strategies considered in the articles of the corpus are reviewed.

Henceforward this article is organized as follows: the CoI model and the three types of presences mentioned in the model are described. In Sect. 3, methodology used in the SR is introduced, particularly the characteristics of the selected studies, the research process aspects and the inclusion and exclusion criteria adopted. Then in Sect. 4, results are presented, a description of the articles included, the didactic-pedagogical strategies adopted by the HEIs since the interruption of face-to-face classes are described, and results from the analysis are showed in the light of the CoI Model dimensions and its effects on the educational community in HEIs, as well as impacts, perceptions and difficulties in the student population, due to the implemented solutions. Section 5 concludes with reflections about the models adopted in HEIs in Ibero-America, their implications and a possible future post-pandemic context.

All these sections are based on the previous article [38], published in the CACIC 2021 Argentine Congress by the same authors, reinforcing the analysis of the documents investigated in the light of the CoI model, and with a more detailed description of the didactic-pedagogical strategies adopted by HEIs since the interruption of the face-to-face activities. Besides, an enriched discussion of the research results is included.

2 Considerations About the CoI Model

The CoI (Community of Inquiry) Model is a theoretical framework developed by Garrison y Anderson [28].

For the online learning in a CoI to be possible, the interrelation of three elements or presences is needed: cognitive, teaching and social [29].

Cognitive Presence. It indicates to what extent students are able to construct meaning through continuous reflection in a critical research community [28, 30], via sustained communication [31]. The proposed model identifies four immutable and non-sequential phases in the cognitive presence: activation, exploration, integration and resolution [28, 30].

Social Presence. It is the capacity of participants to project themselves socially and emotionally and as people, to promote direct communication among individuals and to make the personal representation explicit [32]. Social presence makes the qualitative difference between a collaborative research community and the process of merely uploading information [30].

Teaching Presence. It is defined in the CoI model as the act of designing, facilitating and guiding the teaching-learning processes to obtain the expected results according to students´ necessities and capacities [33]. The relevance of setting a collaborative community to create an effective learning environment comes from the concept of learning based on participation to build and acquire knowledge. The relationship between teachers and students represents the core of the educational experience. It is a complex process in which participants have meaningful and complementary responsibilities. In this model, a series of teachers´ responsibilities are identified, related to the design and organization, thus facilitating the discourse and the direct teaching; they refer to the macrostructure and the process, and to the structural decisions made to adapt to the changes during the educational transaction, facilitating the discourse, aiming at building knowledge [33].

The Internet enables the construction of learning communities, of educational communities of questioning, of inquiry, and of joint creation of knowledge. This possibility is a determining factor for Higher Education (HE) in the 21st century, which must be analyzed to observe its real potential in the development of students' learning [34].

Figure 1 illustrates the categories of each presence, and the indicators involved in the CoI Model.

SOCIAL PRESENCE		
Affection	**Open Communication**	**Cohesion**
To express emotions Humor appeal To express yourself openly	To follow the exposition To quote others' messages To explicitly refer to others' messages To make questions To express regard To express agreement	Vocatives To use inclusive pronouns to talk or refer to the group Phatic elements, greetings

COGNITIVE PRESENCE			
Triggering	**Exploration**	**Integration**	**Resolution**
Evocative (inductive)	Inquisitive (divergent)	Tentative (convergent)	Commited (deductive)

TEACHING PRESENCE		
Educational Design and Organization	**To facilitate the discourse**	**Direct Teaching**
To create a study programme To design methods To establish a timeframe To use the tools effectively To establish rules of conduct and courtesy in the electronic communication (netiquette) To make observations in the macro level of courses content	To identify agreements/ disagreements To try to achieve consensus To encourage, recognize and reinforce students' contributions To create a study environment To request participants' opinions and to promote discussion To assess the process efficacy	To present contents/ issues To focus the debate on specific issues To summarize the debate To confirm what has been understand through assessment and explicative feedback To identify concept mistakes To provide knowledge from different sources; for example, textbooks, articles, Internet, personal experiences To respond to technical concerns

Fig. 1. Dimensions, categories and indicators involved in the CoI Model developed by Garrison y Anderson [28]. **Note.** Adapted from [35].

3   Methodology

3.1   Characteristics of the Selected Studies

All studies were required to inquire about the implementation of solutions that respond to the emergency educational context, due to the interruption of face-to-face classes, considering HE students as population of interest, using Information and Communications Technology (ICT) or Learning and Knowledge Technologies (LKT) and to have been developed in, at least, one country of Ibero-America. From the reading and analysis of the selected corpus, the following questions are proposed to be answered:

Q1: Which are the didactic-pedagogical strategies adopted by the HEIs since the interruption of face-to-face activities due to the COVID-19 pandemic?

Q2: Which were the scope and implications of such strategies on the social, cognitive and teaching presences of the CoI Model?

Q3: Which was the assessment of each presence in the search for solutions before the emergency remote education paradigm?

3.2 Research Strategy

The research period was limited to years between 2020 and 2021, period in which the HEIs decided a partial or total closure, thus making face-to-face education continuity impossible, either temporarily or permanently [3], due to the alarming and uncontrolled spread of the new coronavirus SARS-CoV-2 and its potential illness COVID-19.

A bibliographic and webgraphic review of the published articles was developed, investigating in journals and academic events in Ibero-America, preferentially in Spanish and English, for example, Ibero-American Journal of Education Technology and Technology Education (TE&ET) [Revista Iberoamericana de Tecnología en Educación y Educación en Tecnología], TE&ET Congress, Ibero-American Review on Digital Education [Revista Iberoamericana de Educación a Distancia (RIED)], Virtuality, Education and Science Journal (Argentine UNC) [Revista Virtualidad, Educación y Ciencia (UNC)], Question Journal/Youth (Faculty of Journalism and Social Communication- Argentine UNLP) [Revista Question/Juventudes (FPyCS-UNLP)], Conference on Remote Education (Argentine UNLP) [Jornadas de Educación a Distancia (UNLP)], Institutional Repository of the Argentine University of La Plata [Repositorio Institucional de la Universidad Nacional de La Plata], Intellectual Creation Promotion Service [Servicio de Difusión de la Creación Intelectual (SEDICI)], reports from the Argentine Ministry of Education [Ministerio de Educación de la Nación], State and Public Policies Journal/ Educational Proposal (FLACSO) [Revista Estado y Políticas Públicas/Propuesta Educativa], Journal of the Spanish Institute of Strategic Studies (IEEE-RITA), ACM, Journal Computers & Education (Elsevier), Journal NAER, American Journal of Distance Education (AJDE) and other pertinent ones, using the search engines Google Scholar, eric.ed.gov, worldwidescience.org. y https://dialnet.unirioja.es/. Moreover, studies suggested by experts on the subject were incorporated, and a manual search was carried out on the tables of contents of the Remote Education Journals [revistas de Educación a Distancia (RED)]; Electronic Journal of Educative Investigation [Revista Electrónica de Investigación Educativa (REDIE)] and Scientific Electronic Library Online (SciELO).

This enriched and improved work is a product of the one published by the authors in the CACIC 2021 Congress and is already published in the Minutes Book (http://sedici.unlp.edu.ar/handle/10915/129809).

All the studies include experiences carried out in Latin American countries and Spain. Works were analyzed that, despite having been edited in a European country, included Latin-American issues and studies with regional statistics. Regarding the chain used in the process, a set of terms in Spanish and English were used, as shown in Table 1.

Table 1. Term chain used for the search

	Spanish	English
A1	Educación a distancia	Distance education
A2	Educación virtual	E learning
A3	Aprendizaje electrónico	E learning
B1	Cuarentena	Quarantine
B2	Pandemia	Pandemic
B3	Coronavirus	Coronavirus emergency
C1	Educación superior	Higher education
D1	Interacción	Interaction
E1	Paradigma de comunidad de indagación	Community of inquiry paradigm
E2	Modelo de comunidad de indagación	Community of inquiry model

The search chains correspond to the following Boolean expressions:

(A1 OR A2 OR A3) AND (B1 OR B2 OR B3) AND (C1) AND (D1)

(A1 OR A2 OR A3) AND (B1 OR B2 OR B3) AND (C1) AND (D1) AND (E1 OR E2)

Based on this strategy, documents were selected. They were evaluated in a first stage by reading the title, summary and date. One hundred twenty-seven articles were selected, according to the criterion explained in relation to the research questions proposed in this research and to the inclusion and exclusion criteria mentioned in Table 2, which required the exhaustive and complete reading of the selected corpus. As a result, twenty-four articles were chosen to be presented in this work for analysis and synthesis.

Table 2. Inclusion and exclusion criteria

Inclusion criterion	Exclusion criterion
Empiric studies and theoretical reflections that inquire into the use of VTLE and synchronous and asynchronous tools in educational contexts of HE during the COVID-19 pandemic	Empiric studies and theoretical reflections that DO NOT inquire into the use of VTLE and synchronous and asynchronous tools in educational contexts and DO NOT happen during the period concerned
Empiric studies that inquire into the implementation of solutions and educational models based on ICT and LKT in HE	Empiric studies that DO NOT inquire into the implementation of solutions and educational models based on ICT and LKT in HE
Empiric studies that inquire into techno-educational models, focusing on the CoI one in HE	Empiric studies that DO NOT inquire into techno- educational models, focusing on the CoI one in HE
Studies and research that inquire into the CoI model during the period affected by the COVID-19 pandemic	Studies and research that DO NOT include the CoI model and do not happen during the period affected by the pandemic

4 Results

4.1 Description of the Articles Included in the SR Corpus

Table 3 details the twenty-four selected articles that study the strategies adopted during the interruption of face-to-face activities and that can be analyzed according to the CoI model.

In the selected corpus, eighteen articles include research and six articles are theoretical reflections, two of which are argumentative and testimonial texts, and the rest of them are descriptive analysis and position papers.

From the eighteen Research included in this SR, five of them are surveys and interviews aimed at teachers, two at documental analysis, other four at HEIs and Universities analysis and the rest of them (7) are works that include methodologies of mixed research aimed at students, their experiences, perceptions and emotions during the course in the emergency period.

Research focusing on teachers, [1, 7, 9, 11, 22], are focused on the implemented strategies and the didactic-pedagogical resources used during the emergency; surveys and interviews that analyze the adopted models based on quantitative and qualitative research methodologies. The populations studied range from 26 to 1237 teachers in universities in Ibero-America.

Research including documental analysis methodology, [5, 14], the first of which is developed in a university education scenario in the first semester of 2020, in Peru, considering more than twenty academic texts that include articles, texts and reports related to the university education. The second article is the introduction of a special number of the Review on Digital Education [Revista de Educación a Distancia (RED)], which includes about a dozen and a half articles representing a rich variety of issues and approaches. Authors also represent a diverse cultural background, coming from countries such as Spain, Portugal, Brazil, Mexico and Colombia, besides others from Europe and Asia.

Research studying HEIs [6, 10, 16, 17], include a research with a qualitative approach in a sample of ten HEIs in Colombia, a description of the methodologies adopted in six universities from Ecuador, a summary of the survey by the International Association of Universities that provides interesting data about HE world trends in the context of pandemic, including 424 universities all over the world and finally, a study about web contents from 25 universities from Brazil, Chile, Peru, Mexico and Colombia.

Research involving students from HEIs [4, 12, 13, 15, 19, 21, 23], based on surveys and include descriptive methodologies with mixed analysis in populations with a range from 41 to 548 students of varied careers in universities from Ibero-America.

Texts including theoretical reflections [2, 3, 8, 9, 20, 24], refer to position papers and argumentative reflections based on surveys and testimonies. Some of the main issues addressed in these articles are the tension gap resulting from the adopted models, its implications on students and their families, communication tools that were used and the social imaginary about face-to-face versus emergency remote education.

Works were selected from Mexico (34%), Spain (29%), Peru (13%) Argentina (8%), Cuba (8%), Colombia (4%), Ecuador (4%), as shown in Fig. 2.

Table 3. Selected articles that study adopted strategies. Own elaboration.

	Categories	1	2	3	4	5	6	7	8	9	10	11	12	13	14	15	16	17	18	19	20	21	22	23	24
Social presence	Affection		X		X					X					X							X			
	Communication		X		X					X					X							X			
	Cohesion		X		X					X					X							X			
Cognitive presence	Activation		X		X					X	X	X	X		X									X	
	Exploration		X							X		X	X		X									X	
	Integration		X							X		X	X		X									X	
	Resolution		X							X		X	X		X									X	
Teaching presence	Disign	X	X	X		X	X	X	X	X	X	X	X	X	X	X	X	X	X	X	X		X	X	X
	Discourse		X		X				X	X	X	X	X	X	X	X			X	X		X		X	X
	Teaching		X		X				X	X	X	X	X	X	X	X			X	X		X		X	X

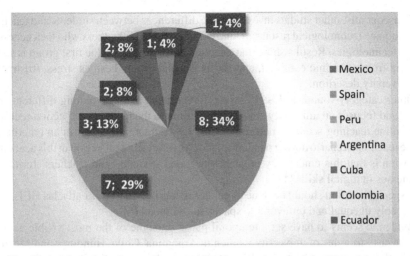

Fig. 2. Articles selected according to publication country of origin. Own elaboration.

4.2   Didactic-Pedagogical Strategies Adopted by HEIs During the Interruption of Face-To-Face Activities.

In order to answer the first research question, the didactic-pedagogical strategies adopted by HEIs in the mentioned period and the actions carried out to deal with the crisis are analyzed, according the selected corpus.

The twenty-four articles analyze or reflect on the consequences that the interruption of the face-to-face HE has had in 2020 due to the COVID-19 pandemic.

This SR made it possible to identify a series of contributions regarding the adopted models, most of them being studies that analyze the didactic-pedagogical resources implemented as a consequence of the interruption of face-to-face classes.

Besides, a case reveals the immediate effects of crisis, what impacts it is having and how the sector is responding to the enormous challenges; it also includes principles on which planification to overcome the crisis should be based [18].

On the other side, other set of articles analyze the gap resulting from the interruption of face-to-face activities, which has not only led to reinvent teaching and to reorganize the teaching-learning process, but it also furthered the real structural conditions of a disadvantaged student population.

Regarding the works that investigate through surveys and interviews to teachers and students, they study these groups´ perception in relation to their skills and competences necessary to integrate learning and communication tools, concluding that a replica of the system was adopted from face-to-face education to remote education, without taking into account the characteristics and the essence of the latter.

During confinement, provisional solutions were provided in non-university levels, most of which will finish once the crisis be overcome, although they will contribute with elements for reflection to adopt certain innovations. However, it will be different at university. Remote, digital, online and flexible modalities are going to be widely used once the crisis pandemic is overcome [8].

Consequently, other studies investigate the differences between students and teachers with adequate technological resources, and compare them with others who lack access to quality technologies. Results show that the main consequence on the first group is stress resulting from academic overload, and for the second group, besides stress, frustration and university desertion.

Other category found in this SR is the comparative analysis among different scenarios, and focuses on university students in relation to the instructional action received in an online teaching scenario before the COVID-19 quarantine, and in an emergency remote teaching scenario during pandemic [15]. An interesting reflection in this academic dimension is that this emergency crisis has revealed students´ and teachers´ formative weaknesses in digital skills [11].

A special mention should be done of gender issues in HE. Historically, las HEIs have been deeply unequal and unfavorable spaces for women [16].

Pandemic seems to have set a temporal pause on some of the most visible actions of the strong struggles of university women in Argentina, Chile, Honduras and Mexico, among other countries. Moreover, the first reports and analysis about remote academic activities show a worsening of gender inequalities with an overload of domestic and care activities [36].

In sum, all the adopted models that were analyzed in this bibliography review reveal innumerable deficiencies aggravated by the interruption of face-to-face classes in HEIs due to the pandemic.

The differences between students and teachers regarding the skills to use digital tools, Internet access, and lack of adequate devises to course show a gap that will be accentuated, according to several authors, in the post-pandemic.

In most of the selected works, a future context of hybridization is foreseen, which tense the necessary skills and the Access to the new ICT and LKT in the educational communities of the HEIs.

All the reflections and conclusions of this selected corpus result in the need to train teachers in pursuit of the new scenarios and policies that allow students access to the new technologies.

Most of the adopted models studied in this SR include synchronous and asynchronous communication tools, mediated by virtual environment that has been the solution to the interruption of face-to-face classes. The pedagogical models that have replaced the face-to-face teaching in an emergency remote context were implemented at a time when the remote teaching negative assessment is explained by the perceived inverse relationship between dedication to study and academic performance, and by the lack of adaptation of teachers to students´ personal and academic circumstances. In summary, it is affirmed that university must turn into more collaborative and student-centered models [19].

4.3 SR Analysis Based on the Community of Inquiry

Related to the second and third research questions, the adopted models are assessed and results obtained according the CoI model are shown.

In the twenty-four selected articles, the Teaching presence dimension is taken into account, especially the category educational design and organization. Such category refers to design and organization problems including macrostructural decisions adopted before the beginning of the teaching process, and those adopted to adapt oneself to changes during the educational transaction [33].

Within the dimension mentioned before, fifteen articles that deal with the categories facilitating the discourse and direct teaching in the CoI model are studied [28].

The latter articles study the tools and means implemented with a didactic-pedagogical approach, thus facilitating discourse and teaching, providing information, guiding learning, developing skills, motivating and calling students´ interest, and also creating expression, simulation and content production environments [24].

In the selected corpus in this SR, only eight articles deal with the Cognitive presence, with emphasis on the category Activation, which is recognized as a triggering event, followed by deliberations, conception and action guarantee [33].

With respect to the Social presence dimension, only five articles of this SR focus on what consists in three indicators in the CoI model: affective communication, open communication and group cohesion, which must be dealt with in the instructional design as well as in the teaching-learning process development [37].

This last group is composed of theoretical reflections and a descriptive analysis that analyze collaborative work strategies, methodologies that put emphasis on the group cohesion and the importance of the affective communication in the teaching-learning processes in a virtual learning environment (VTLE). The reflections and results of these works coincide with the difficulties found by students to communicate with mates, to deal with an online collaborative learning approach, to recognize the importance of the group cohesion that integrate the educational community in the learning process, and how to learn to work collaboratively in virtuality, especially in an emergency context.

In Fig. 3, it can be identified that research of the selected corpus deal with the emergency problem from the design and organization processes perspective, present in the Teaching presence dimension, and to a lesser extent, the categories that intervene in Cognitive presence and by far less emphasis on the categories of the Social presence dimension.

5 Discussion and Conclusion

Research can be identified in the works included in the review corpus; most of which focuses on the Teaching presence dimension, and particularly on educational design and organization category.

A high percentage of the works included in this article investigate the synchronous and asynchronous tools selected by teachers to interact with students, the strategies proposed for the online class and the lack of technological resources such as students´ and teachers´ deficiency of the Internet connection in the countries involved in the study.

Other works deal with the students´ and teachers´ training and experiencing problem to use technological devises and communicative tools, taking into account that institutions had a VTLE. The vast majority of the comparative studies that relate the virtual education experience and compare it with emergency remote education, its implications in the students´ motivation and the pedagogical activities design, conclude in the negative perspective of these technological innovations used to face the crisis.

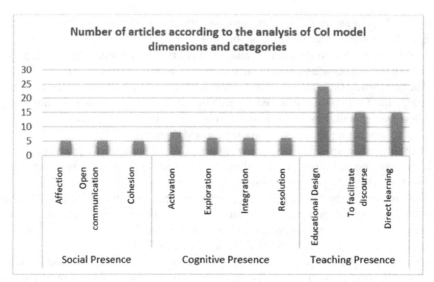

Fig. 3. SR analysis based on the CoI model. Articles selected according to dimensions and categories. Own elaboration.

To a lesser extent, the reviewed works deal with the Cognitive presence, in terms of how the different stages of the performance of the educational proposal create meaning and deal with contents from a critical perspective. However, those who analyze this aspect emphasize the difference on teachers´ and students´ digital training, stressing the quality of education in this context [38].

The research included in this SR evidence that the scarce production on the Social presence dimension and the analysis focused on the affection, group cohesion and open communication, which are indicators reinforcing learning and supporting a dynamic of positive social relationship. This aspect is only treated in some articles from the perspective of cohesion among first-year students in the HEIs, and as a desertion variable, considering the high schooling of students who has passed through high school.

On the other side, some articles make recommendations on tools of collaborative work and synchronous communication tools for students´ interaction but, at the same time, they report the deficiencies and the fatigue resulting from these means of communication due to the lack of resources that makes it impossible for the students to be able to express himself openly, to follow the exposition or to expressly refer to others´ messages.

It is evidenced, according to the research, the need to rethink the didactic-pedagogical strategies implemented by universities and HEIs, in a VTLE, in relation to the Social presence dimension of the CoI model, with a view to a post-pandemic context, where many of these methodologies adopted during the confinement period will remain in force [38].

The derivations of this study provide a reference frame for future works that investigate experiences carried out by universities and HEIs and their implications, in a future context of post-pandemic blended proposals as well as the strategies that favor interactions and the collaborative teaching-learning processes mediated by virtual environments [38].

The conclusions of this articles will be considered in the process of advancing the thesis "Study of the Interactions between Teachers and Students in Digital Environments within the framework of Pedagogical Continuity in Quarantine Context" to achieve the Master´s Degree in Application of Computer Technology in Education, by the School of Computer Sciences of the Argentine University of La Plata. Besides, they allow for further analysis on the design of educational proposals based on the three dimensions, to consider their impacts on the educational process [38].

References

1. Amaya, A., Cantú, D., Marreros, J.G.: Análisis de las competencias didácticas virtuales en la impartición de clases universitarias en línea, durante contingencia del COVID-19. RED. Revista Educación a Distancia **21**(64) (2021). https://doi.org/10.6018/red.426371
2. Cabero-Almenara, J., Llorente-Cejudo, C.: COVID-19: transformación radical de la digitalización en las instituciones universitarias. Campus Virtuales **9**(2), 25–34 (2020). (www.revistacampusvirtuales.es)
3. Canaza-Choque, F.A.: Educación Superior en la cuarentena global: disrupciones y transiciones. Revista Digital De Investigación en Docencia Universitaria **14**(2), 1–10 (2020)
4. Cano, S., Collazos, C.A., Flórez-Aristizabal, L., Moreira, F., Ramírez, M.: Experiencia del aprendizaje de la Educación Superior ante los cambios a nivel mundial a causa del COVID-19. Campus Virtuales **9**(2), 51–59 (2020). www.revistacampusvirtuales.es
5. Chiparra, W.E.M., Vasquez, K.M.C., Casco, R.J.E., Pajuelo, M.L.T., Jaramillo-Alejos, P.J., Morillo-Flores, J.: Disruption caused by the COVID-19 pandemic in Peruvian University education. Int. J. High. Educ. **9**(9), 80–85 (2020)
6. Díaz-Guillen, P.A., Andrade Arango, Y., Hincapié Zuleta, A.M., Uribe, A.P.: Análisis del proceso metodológico en programas de educación superior en modalidad virtual. RED. Revista Educación a Distancia **21**(65) (2021). https://doi.org/10.6018/red.450711
7. Expósito, C.D., Marsollier, R.G.: Virtualidad y educación en tiempos de COVID-19. Un estudio empírico en Argentina (2020)
8. Aretio, L.G.: COVID-19 y educación a distancia digital: preconfinamiento, confinamiento y posconfinamiento. RIED. Revista Iberoamericana de Educación a Distancia **24**(1), 09–32 (2021). https://doi.org/10.5944/ried.24.1.28080
9. Gómez, Á.I.P.: Capítulo II Repensar el sentido de la educación en tiempos de pandemia. La formación del pensamiento práctico, el cultivo de la sabiduría. Reconstruyendo la educación superior a partir de la pandemia por COVID-19, **32**
10. Toala, J.M.I, Tigua., M.X.L., Farfán, F.A.L., Jaime, L.P.M.: Educación virtual una alternativa en la educación superior ante la pandemia del covid-19 en Manabí. UNESUM-Ciencias. Revista Científica Multidisciplinaria **5**(1), 1–14 (2020). ISSN 2602–8166, https://doi.org/10.47230/unesum-ciencias.v5.n1.2021.328
11. Gallegos, J.C.P., de la Torre B.A.T., Sprock, A.S.: Distance education. An emerging strategy for education in the pandemic COVID-19. In: 2020 XV Conferencia Latinoamericana de Tecnologias de Aprendizaje (LACLO), pp. 1–11 (2020). https://doi.org/10.1109/LACLO50806.2020.9381137
12. Lovón, M., Cisneros, S.: Repercusiones de las clases virtuales en los estudiantes universitarios en el contexto de la cuarentena por COVID-19: El caso de la PUCP. Propósitos y Representaciones **8**(SPE3), e588 (2020).https://doi.org/10.20511/pyr2020.v8nSPE3.588

13. Martin, M.V., Vestfrid, P.: Reinventar la enseñanza en tiempos de COVID-19. In: III Jornadas sobre las Prácticas Docentes en la Universidad Pública (Edición en línea, junio de 2020) (2020)

14. Teixeira, M.A., Zapata-Ros, M.: Introducción/presentación al número especial de RED Transición de la educación convencional a la educación y al aprendizaje en línea, como consecuencia del COVID-19. Revista Educación a Distancia (RED) X(X) (2021). https://doi.org/10.6018/red.462271

15. Niño, S., Castellanos-Ramírez, J.C., Patrón, F.: Contraste de experiencias de estudiantes universitarios en dos escenarios educativos: enseñanza en línea vs. enseñanza remota de emergencia. Revista Educación a Distancia (RED) 21(65) (2021). https://doi.org/10.6018/red.440731

16. Ordorika, I.: Pandemia y educación superior. Revista de la educación superior 49(194), 1–8 (2020)

17. Paredes-Chacín, A., Inciarte, A., Walles-Peñaloza, D.: Educación superior e investigación en Latinoamérica: transición al uso de tecnologías digitales por COVID-19. Revista de Ciencias Sociales (Ve) XXVI(3), 98–117 (2020)

18. Pedró, F.: COVID-19 y educación superior en América Latina y el Caribe: efectos, impactos y recomendaciones políticas. Análisis Carolina 36(1), 1–15 (2020)

19. López, E.P., Atochero, A.V., Rivero, S.C.: Educación a distancia en tiempos de COVID-19: análisis desde la perspectiva de los estudiantes universitarios. RIED. Revista Iberoamericana de Educación a Distancia 24(1), 331 (2020). https://doi.org/10.5944/ried.24.1.27855

20. Ramos, M.L.R., Ruelas, M.R.: Capítulo 6 La realidad de las personas con discapacidad frente a la COVID-19 en educación superior. La pandemia de la COVID-19–19 como oportunidad para repensar la educación superior en México 119

21. Roig-Vila, R., Urrea-Solano, M., Merma-Molina, G.: La comunicación en el aula universitaria en el contexto del COVID-19 a partir de la videoconferencia con Google Meet. RIED. Revista Iberoamericana de Educación a Distancia 24(1), 197–220 (2021). https://doi.org/10.5944/ried.24.1.27519

22. Mendiola, M.S., et al.: Retos educativos durante la pandemia de COVID-19: una encuesta a profesores de la unam. Revista Digital Universitaria (RDU) 21(3) (2020). https://doi.org/10.22201/codeic.16076079e.2020.v21n3.a12

23. Sureima, C.F., et al.: El aula virtual como entorno virtual de aprendizaje durante la pandemia de COVID-19. In: aniversariocimeq2021, March 2021

24. Vidal, M.N.V.: Estrategias didácticas para la virtualización del proceso enseñanza aprendizaje en tiempos de COVID-19. Educ. Méd. Superior 34(3) (2020)

25. Naciones Unidas [NU]. COVID-19 y educación superior: El camino a seguir después de la pandemia (2020). Recuperado de https://www.un.org/es/impacto-académico/ COVID-19 y-educación-superior-el- camino-seguir-después-de-la-pandemia

26. IESALC, «Global University Network for Innovation. Publications. Report "COVID-19 y educación superior: De los efectos inmediatos al día después. Análisis de impactos, respuestas políticas y recomendaciones",» 6 abril 2020. [En línea]. http://www.guninetwork.org/files/COVID-19-060420-es-2.pdf. Último acceso: 05 julio 2021

27. Hodges, C.B., Moore, S., Lockee, B.B., Trust, T., Bond, M.A.: The difference between emergency remote teaching and online learning. 27 Mar 2020, Educause Review. https://bit.ly/34tYI9r

28. Garrison, D.R., Anderson, T.: E–Learning in 21st Century: A Framework for Research and Practice. Routledge Falmer, London (2003)

29. Garrison, D.R., Cleveland-Innes, M., Fung, T.S.: Exploring causal relationships among teaching, cognitive and social presence: student perceptions of the community of inquiry framework. Internet High. Educ. 13(1–2), 31–36 (2010). https://doi.org/10.1016/j.iheduc.2009.10.002

30. Garrison, R., Anderson, T., Archer, W.: Critical inquiry in a text based environment: computer conferencing in higher education. Internet High. Educ. **11**(2), 1–14 (2000). https://doi.org/ 10.1016/S1096-7516(00)00016-6

31. Gunawardena, C.N., Lowe, C.E., Anderson, T.: Analysis of a global online debate and the development of an interaction analysis model for examining social construction of knowledge in computer conferencing. J. Educ. Comput. Res. **17**(4), 397–431 (1997)

32. Akyol, Z., Garrison, D.R., Ozden, M.Y.: Online and blended communities of inquiry: exploring the developmental and perceptional differences. Int. Rev. Res. Open Dist. Learn. **10**(6), 65–83 (2009)

33. Gutiérrez-Santiuste, E., Rodríguez-Sabiote, C., Gallego-Arrufat, M.-J.: Cognitive presence through social and teaching presence in communities of inquiry: a correlational–predictive study. Australas. J. Educ. Technol. **31**(3), 349–362 (2015)

34. Gutiérrez-Santiuste, E., Gallego-Arrufat, M.J.: Presencia social en un ambiente colaborativo virtual de aprendizaje: análisis de una comunidad orientada a la indagación. Revista mexicana de investigación educativa **22**(75), 1169–1186 (2017). Recuperado en 05 de julio de 2021, de http://www.scielo.org.mx/scielo.php?script=sci_arttext&pid=S1405-666620170 00401169&lng=es&tlng=es

35. Sánchez, M.R.F., Berrocoso, J.V.: A community of practice: an intervention model based on computer supported collaborative learning. Comunicar. Media Educ. Res. J. **22**(1) (2014)

36. Hupkau, C., Petrongolo, B.: Work, care and gender during the COVID-19 crisis (2020). Accessed from London: https://cieg.unam.mx/COVID-19-genero/pdf/reflexiones/academia/ work-care-and-gender.pdf

37. Hughes, M., Ventura, S., Dando, M.: Assessing social presence in online discussion groups: a replication study. Innov. Educ. Teach. Int. **44**(1), 17–29 (2007). https://doi.org/10.1080/147 03290601090366

38. Spandre, O., Dieser, P., Sanz, C.: Revisión sistemática de metodologías educativas implementadas durante la pandemia por COVID-19 en la Educación Superior en Iberoamérica (2021). http://sedici.unlp.edu.ar/handle/10915/129809

Software Tool for Thematic Evolution Analysis of Scientific Publications in Spanish

Santiago Bianco[1]([✉]), Laura Lanzarini[2], and Alejandra Zangara[2]

[1] Information Systems Research Group, UNLa (GISI-UNLa),
Buenos Aires, Argentina
sabianco@unla.edu.ar
[2] Computer Science Research Institute LIDI (III-LIDI) (UNLP-CICPBA),
Buenos Aires, Argentina
{laural,azangara}@lidi.info.unlp.edu.ar

Abstract. Retrieving information from text documents on a specific theme is of great interest in different areas. A difficulty that is always present when documents from different periods are analyzed is the thematic evolution, an aspect that impacts the quality and results of the searches carried out. Identifying changes in terminology and the evolution in study themes is of great interest to disciplines such as bibliometrics and scientometrics. This article describes the development of a software tool capable of implementing the steps of the methodology suggested in [2] and [3]. Using various bibliometric indicators, this methodology can analyze thematic evolution in scientific documents. In addition, the software tool proposed here will allow the user to experiment with new metrics and modifications to the methodology more easily by applying incremental iterative development. The tests carried out allow us stating that the inclusion index is an adequate metric to select the most relevant theme relationships, facilitating the understanding and visualization of the results obtained.

Keywords: Bibliometric analysis · Text mining · Thematic evolution

1 Introduction

The analysis of text documents from specific contexts is a theme of interest for different areas such as information retrieval, document classification, bibliometric and scientometric analysis, and so forth.

When processing text documents written in different time periods, thematic evolution is an aspect that must be taken into account. Identifying the changes that took place over time in the nomenclature used on various themes within the same discipline or discourse area is an extremely useful tool when trying to apply text mining strategies.

As a specific case, any teacher, researcher or student who needs to write an article, thesis or research work, will have to carry out a review of the corresponding state of the art. In this sense, the possible themes of interest within a

© The Author(s), under exclusive license to Springer Nature Switzerland AG 2022
P. Pesado and G. Gil (Eds.): CACIC 2021, CCIS 1584, pp. 64–77, 2022.
https://doi.org/10.1007/978-3-031-05903-2_5

particular domain have to be identified. This bibliographic search process is generally time-consuming and, if not oriented correctly, can lead to blockages and frustration for the researcher. It would be interesting then to have methods and tools available to simplify the search and analysis of bibliography or publications of any kind for these processes.

In the first instance, bibliometric tools could be used to carry out the initial analysis of the texts of interest. Bibliometrics is known as a discipline capable of describing a set of publications applying statistical analysis techniques, identifying relevant themes focuses, author collaboration networks, information on citations, and so forth. Scientometrics is a sub-discipline of bibliometrics that focuses specifically on scientific publications.

In any case, these approaches generally allow quantitative analyzes such as a list of the most cited authors, institutions with the highest number of publications, themes most written about, and so forth. When a more in-depth qualitative analysis is required, text mining and visualization techniques should be applied, such as thematic maps together with traditional bibliometric methods.

Thematic maps are a way of representing different themes covered in a field of a scientific discipline at a certain moment. Different types of bibliometric information can be used to build these graphs, one of them being the analysis and correlation between relevant terms.

An analysis technique called thematic evolution is derived from the thematic maps. It consists of showing the "evolution" of the relevance of a particular theme on a timeline. For example, it could be shown that in 2010 there was a theme focus dedicated to research in neural networks and that the same group of people who worked on this theme gradually moved their research interests towards a different theme, such as black box model interpretability. The idea is to show that the first theme mutated or evolved into the second one. It should be noted that, in this context, the term "evolution" means "change and transformation", and does not imply there has been an improvement. Thus, saying that "Theme A evolved into Theme B" does not necessarily mean that Theme B is better in some ways than Theme A.

Even though there are some tools that allow these analyses, very few are available for the Spanish language, and they are not so intuitive that they can be used by non-expert users in the computer area. In addition, they are aimed at scientific publications in English downloaded from portals such as Web of Science or Scopus, with a particular format.

In this context, this article describes a tool that allows applying the methodology proposed in [3] together with the results obtained through its use.

2 Proposed Methodology

2.1 Gathering the Documents

To carry out a thematic analysis, documents that are representative of the area of interest over a period of time long enough to be able to divide it into sub-periods (small periods of time into which a larger interval is divided) are required. Then, using thematic evolution, relationships between the central themes from different sub-periods will try to be identified.

When accessing the documents, we decided to work with scientific journals that would use the Open Journal System (OJS) for their digital issues, which allows consulting all available articles in a systematic and repeatable way, as they are supported on the same standard system. OJS [8] is an open source solution for managing and publishing academic journals online, thus reducing publication costs compared to print versions and other forms of dissemination. Unlike Scopus, Web of Science and others, no access codes or any other type of authentication are required to extract information from the platform. This allows automating the journal articles extraction process through a script, which can easily be modified to extract the data from any journal that is implemented on OJS.

Raw data are downloaded as plain text. Key elements such as title, year of publication, abstract, and author's address are automatically pulled from the OJS system, without the need to directly access the full article. Authors and country affiliations are identified from their addresses and available metadata.

On the other hand, document publication language must be selected in advance. In this article, emphasis is placed on scientific documents in Spanish because this is still a little studied language from the point of view of thematic evolution. Inconsistent expressions, special characters, and ambiguities are processed after their download and collection, in a separate script. This script is used to give the final format to the publications so that they can be input to the algorithms used in the analysis.

2.2   Document Analysis

To analyze the documents collected, the title, abstract and keywords indicated by the authors are used. All these sections must be preprocessed and grouped so that they can be used properly by the algorithms. The process consists of the following steps:

1. The terms contained in the previously mentioned sections (title, abstract and keywords indicated by the authors) are extracted and standardized. This normalization process consists of replacing special characters, unifying synonyms and acronyms, and writing all terms in lowercase.
2. The n-grams obtained, made up of two, three or four words, are added to the set of terms.
3. Those that exceed a certain threshold value of TD-IDF (Term Frequency-Inverse Document Frequency) are selected from the set of terms. This metric assigns high values to those terms that have a high frequency (in the given document) with a low frequency in the entire collection of documents, thus filtering common terms [10].
4. All selected terms are unified in a corpus to be analyzed by the algorithm.

2.3 Research Themes Identification

To identify the research themes and/or centers of interest for researchers, the joint occurrence or co-occurrence of previously identified terms is used [5]. This co-occurrence is calculated as indicated in 1, where c_{ij} is the number of documents in which both terms appear together, and c_i and c_j are the number of documents in which they appear individually.

$$e_{ij} = \frac{c_{ij}}{c_i c_j} \tag{1}$$

Using these co-occurrence values, the simple center algorithm [7] is applied to build thematic networks made up of subgroups of strongly linked terms that correspond to interests or research issues of great importance in the academic field.

The detected networks can be represented using the density and centrality measures defined in [4].

Centrality measures the degree of interaction between networks. Its value is calculated as indicated in Eq. 2, where k is a keyword belonging to the theme and h is a keyword belonging to other themes. It represents the strength of external links with other themes, and can be considered a measure of the importance of a theme in the development of the entire research field being analyzed.

$$c = 10 \sum e_{kh} \tag{2}$$

Density measures the internal strength of the network, and its value is calculated as indicated in Eq. 3, where i and j are the keywords belonging to the theme and w is the number of keywords in the theme. Density measures the strength of the internal links between all the keywords that describe the research theme. This value can be understood as a measure of the development of the theme.

$$d = 100 \left(\sum e_{ij}/w \right) \tag{3}$$

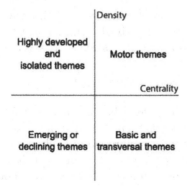

Fig. 1. Quadrant distribution in a thematic map [6]

Based on these two measures, research themes can be represented in a two-dimensional strategy diagram with four quadrants, ordered by centrality and density, as shown in Fig. 1. The themes in the upper right quadrant, known as motor themes, are generally well developed and important in structuring a field of inquiry. The themes in the upper left quadrant are well developed but have marginal importance for the field, as they have a lot of interrelation with each other but little with the rest of the themes in the other quadrants. The themes in the lower left quadrant are weakly related and poorly developed, indicating that they are emerging or declining themes, on the way to disappearing. Finally, in the lower right quadrant are the themes that are relevant to some fields of research but are poorly developed. The latter includes basic, transversal and general themes.

By analyzing the relationship between the terms that make up the different thematic networks within a discipline in different sub-periods of time, the development of a given theme over the years can be analyzed, and the changes in relevant theme focuses can be seen. This is known as thematic evolution, and is discussed in more detail in the next section.

2.4   Thematic Evolution

A thematic area is a set of themes that have evolved over different sub-periods. Each theme is made up of a set of terms. Let T_t be the set of themes detected in sub-period t and let $U \epsilon T_t$ be a theme detected in sub-period t. Let $V \epsilon T_{t+1}$ be a theme detected in the following sub-period $t + 1$. A thematic evolution from theme U to theme V is considered to have happened if there are common terms in both sets. Each $k \epsilon U \cap V$ term is considered a *thematic link*. To weight the importance of a thematic link, the inclusion index defined in [11] calculated according to Eq. (4) is used. This index is a simple metric that in this context is used to measure how strong the relationship between two themes is. Its value is between 0 and 1; a higher value corresponds to a stronger relationship.

$$inclusion = \frac{\#(U \cap V)}{min(\#U, \#V)} \tag{4}$$

If a theme from a sub-period has no thematic link with another theme from a later sub-period, it is considered to be discontinuous, whereas if there is a theme unrelated to a previous sub-period, it is considered as a new or emerging theme.

3   Features of the Software Tool

Even though the methodology described in the previous sections can be applied to any type of text from various thematic areas, it requires at least knowledge of statistics and text mining. This can present difficulties for teachers, researchers or those interested in using it who are experts in other disciplines. For this reason, this article proposes the development of a software tool that allows not only automating the process, but also facilitating the execution of the methodology described in the previous section by non-expert users.

3.1   Development Methodology

The development of the software tool is based on an evolutionary prototyped methodology.

A prototyped methodology allows small incremental iterations to be made in the product until the final software is generated. This not only facilitates its modification after usability and validation tests with experts, but also allows the expansion of the tool as the investigation progresses, adding new metrics or processes that improve the results.

Due to the usability requirements of the tool, it must be validated by users to verify that no specific knowledge is required to use it. In addition, experts are required to validate the final results obtained through its use. It is then likely that modifications will be required after these validations, so it is considered that the chosen methodology is appropriate.

3.2   Functional Requirements

The tool should be capable of applying all of the steps of the methodology detailed in [3], together with the visualization improvements proposed there. Specifically, it should be capable of:

- Downloading journal data in OJS format and format them appropriately
- Loading journal data already formatted to be analyzed by the tool
- Automatically processing and filtering missing data or outliers
- Processing article abstracts and adding them to the keywords to be analyzed
- Generating Sankey networks corresponding to the thematic evolution of the texts
- Showing the thematic maps generated for each sub-period analyzed for the thematic evolution
- Filtering the thematic evolution graph according to the inclusion index of the thematic links found

3.3   Non-Functional Requirements

A fundamental requirement for the tool is that it can be used by non-expert users. This implies automating methodology application as much as possible and minimizing algorithm parameterization. Ideally, the user would only have to choose the cutoff years for the sub-periods to be analyzed, and the software would be in charge of choosing the best configuration among the possible parameters to be used. These parameters are:

- Maximum number of terms to be considered in each thematic cluster
- Minimum frequency of a term to be considered significant within a theme
- Maximum size of n-grams to use

In principle, a grid search strategy is used, aimed at maximizing the inclusion index in the relationships identified between terms, although other methods - such as Bayesian optimization and genetic algorithms [1]- will also be tested. Finally, the interface should be simple and direct, explaining the results obtained clearly without applying formulas or implementation details.

4   Results

This section details the results obtained after performing the first iteration of the aforementioned software tool. To test the developed tool, the analysis carried out in [3], was replicated, in which documents from the *EDUTEC* journal were used. This journal was selected because it addresses two specific themes; namely, technology applied to education and technology education. The latter is relevant because all publications use a limited vocabulary. In addition, it has publications in Spanish and has numbers published for more than 25 years.

The set of documents obtained was divided into three sub-periods of similar duration:

– Sub-period 1: 1995–2005
– Sub-period 2: 2006–2013
– Sub-period 3: 2014–2020

For each sub-period, the most relevant themes were identified using two different strategies - first, using only the keywords and second, adding abstracts and titles to these keywords. In each case, the relationships identified were considered and the most relevant ones were selected according to their inclusion level value. For this, the developed tool allows easily running any of the two strategies, and it can also apply dynamic filtering to the results obtained according to the inclusion index of the relationships found between themes.

For all tests, the parameters described in Table 1 were used for thematic evolution analysis, obtained using this tool.

Table 1. Analysis parameters for the *EDUTEC* journal

Parameter description	Value
Maximum number of terms to consider in each thematic cluster	250
Minimum frequency for a term to be considered significant within a theme	20
Maximum size of n-grams to use	3

As a result of the first strategy, that is, the thematic evolution from sets of terms selected taking into account only keywords, the graph in Fig. 2 was obtained. In this figure, the thickness of the bands that join the themes in the different sub-periods is proportional to their inclusion index. In other words, the wider the band that joins two themes, the greater the value of the inclusion index between the two. As it can be seen, despite the fact that the number of terms involved is scarce, it is somewhat complex to identify the most relevant ones.

For example, a simpler and more readable visualization applying a filter with a threshold of 0.5 for the inclusion index can be observed in Fig. 3. By eliminating

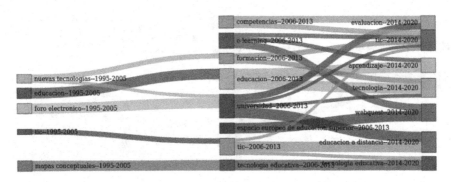

Fig. 2. Thematic evolution analysis using only keywords

the themes with lower inclusion, charts become easier to read and the possibility of considering unreliable results in the analysis is reduced. Thus, it is easier to visualize those terms whose relationship between sub-periods is supported by a higher level of inclusion.

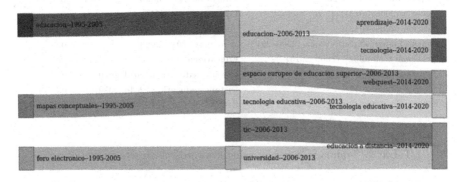

Fig. 3. Thematic evolution results using only keywords, filtered by inclusion index

Applying the second strategy, as expected, the themes identified were more closely related to the documents. This is reflected in how themes are linked, shown in Fig. 4. This figure shows the relationships in a clearer way in the absence of an excessive number of crosses between connections, as it was the case in Fig. 2. Regardless of this, the value of the inclusion index for each pair of terms is still directly proportional to the importance of their relationship.

Table 2 shows the highest inclusion level values obtained as a result of the first procedure. These values correspond to the most interrelated themes, which were graphically joined with the thickest bands in Fig. 2.

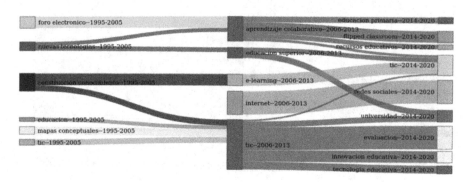

Fig. 4. Thematic evolution using keywords and terms from abstracts and titles

Table 2. Summary of results obtained using keywords only.

Theme A (Sub-period)	Theme B (Sub-period)	Inclusion
Education (1995–2005)	Education (2006–2013)	0.5
Electronic forum (1995–2005)	University (2006–2013)	0.5
Conceptual maps (1995–2005)	Educational technology (2006–2013)	0.5
Education (2006–2013)	Learning (2014–2020)	0.5
Education (2006–2013)	Technology (2014–2020)	0.5
European higher education area (2006–2013)	Webquest (2014–2020)	0.5
Educational technology (2006–2013)	Educational technology (2014–2020)	0.5
ICTs (2006–2013)	Distance education (2014–2020)	0.5
University (2006–2013)	Distance education (2014–2020)	0.5
Competencies (2006–2013)	Evaluation (2014–2020)	0.33

Table 3. Summary of results using keywords, abstracts and titles.

Theme A (Sub-period)	Theme B (Sub-period)	Inclusion
Internet (2006–2013)	Social media (2014–2020)	1
ICTs (2006–2013)	Evaluation (2014–2020)	1
Knowledge-Building (1995–2005)	E-Learning (2006–2013)	0.5
Electronic forum (1995–2005)	Collaborative learning (2006–2013)	0.5
Collaborative learning (2006–2013)	Flipped classroom (2014–2020)	0.5
E-Learning (2006–2013)	ICTs (2014–2020)	0.5
ICTs (2006–2013)	Educational innovation (2014–2020)	0.5
Conceptual Maps (1995–2005)	ICTs (2006–2013)	0.33
ICTs (2006–2013)	Educational technology (2014–2020)	0.33
Knowledge-Building (1995–2005)	ICTs (2006–2013)	0.25

Table 3 summarizes the results obtained with the second procedure; i.e., using n-grams of abstracts and titles in the analysis. As it can be seen, the first terms have high inclusion level values, indicating a consolidated relationship between both periods. Additionally, because new terms are added to represent document content, new themes appear, such as social media, flipped classroom and e-learning.

Through the expert consultation method, it was determined that the results obtained by this method are related to the development of ideas about teaching and ICTs in the field of research.

For example, the themes related to "Electronic Forums" may have become "Collaborative Work," since the ideas of collaborative work and learning went through an automation process when tools became available to carry out tasks. Thus, it would make sense for articles that covered tools to start working on conceptual models. This may also explain the transition between "Collaborative Learning" and "Flipped Classroom".

The transition between "ICTs" and "Educational Technology" could also be understood if we consider the discipline that builds conceptual models that include the use of technological tools. Due to the development methodology that is being applied, if negative feedback is received from the experts, the tool would have to be adjusted by modifying its key aspects, until satisfactory results are achieved: parameter selection, visualization metrics, and algorithms used in data pre-processing.

As additional results to those obtained in [3], the tool allowed generating, together with the thematic evolution network, the various thematic maps corresponding to each analyzed superperiod. As an example, Fig. 5 shows the map corresponding to the 2006–2013 subperiod. The following groups are shown:

- Motor themes: higher education, Internet
- Highly developed and isolated themes: e-learning
- Emerging or declining themes: interaction, collaborative learning
- Basic and transversal themes: ICTs

he results obtained in [2] were also expanded by replicating the analysis previously carried out with articles from the $TEyET$ journal, and adding articles from the end of 2020 and 2021. The thematic evolution analysis was carried out including information from the abstracts and using the following sub-periods, similar to those used in the previous analysis:

- Sub-period 1: 2006–2013
- Sub-period 2: 2014–2018
- Sub-period 3: 2019–2021

In this case, the parameters described in Table 4 were used, obtaining the results represented in Fig. 6. The most remarkable aspect about this is the emergence of the "pandemic" theme in the last sub-period, where many theme groups from other sub-periods converge. This would indicate that there was a change of

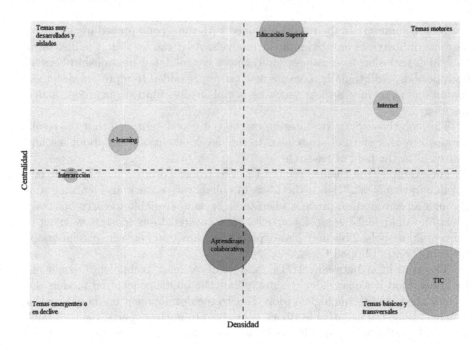

Fig. 5. Thematic map from the 2006–2013 sub-period

Table 4. Parameters used for the analysis of $TEyET$ journal

Parameter description	Value
Maximum number of terms to consider in each thematic cluster	250
Minimum frequency for a term to be considered significant within a theme	20
Maximum size of n-grams to use	3

focus in the themes covered in the publications from this journal, understandable due to the global phenomenon experienced during this period that certainly impacted the disciplines of education and technology.

To focus more on this sub-period, the tool can be used to analyze in more detail the corresponding thematic map. This is shown in Fig. 7, where the following groups can be observed:

– Motor themes: collaborative work, collaborative learning
– Highly developed and isolated themes: mobile devices, serious games, higher education
– Emerging or declining themes: engineering, tangible interaction
– Basic and transversal themes: didactics, education, pandemic

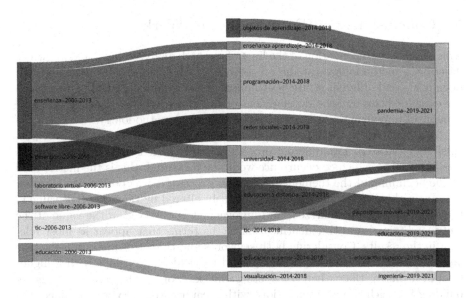

Fig. 6. Thematic evolution analysis results for the *TEyET* journal

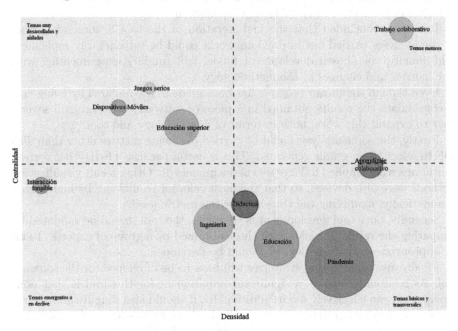

Fig. 7. Thematic map corresponding to the 2019–2021 supberiod, of the TEyET journal analysis

5   Conclusions and Future Lines of Work

This article is a continuation of the lines of research analyzed in [2] and [3], which describe a methodology capable of analyzing thematic evolution in scientific documents and different strategies to improve the results obtained. These works are expanded through the development of the first prototype of a software tool capable of implementing the steps of the suggested methodology.

As a first test to validate the tool, the analyses carried out in [3] were successfully replicated. Two different procedures were carried out when creating the sets of terms on which co-occurrence would be measured; this is the metric used at the beginning of the theme identification process. It was possible to corroborate again that, by adding n-grams previously filtered by their TD-IDF value in the set of documents analyzed, an improvement is observed in the value of the metrics obtained and the identification of the terms that appear as most relevant in the results. Considering both keywords and the terms present in the title and the abstract of each document proved to yield relationships with a higher level of inclusion than when considering only keywords. This is because theme building is enriched and intersections with greater cardinality are obtained. On the other hand, both procedures proved that filtering the relationships by level of inclusion is effective in simplifying visualization.

It is then concluded that the first iteration of the tool is successful, since all the analyses carried out in previous works could be satisfactorily replicated. Additionally, the chosen development model will simplify experimenting with new metrics and changes to the methodology.

Even though significant progress has been made in the research by being able to consolidate the results obtained in a piece of software, there are still several ways to expand this work, both in terms of methodology and tool.

Firstly, the same analysis could be carried out using metrics other than TF-IDF to select the n-grams generated. This is useful because TF-IDF has certain limitations when applied to large sets of documents [9]. Other result visualization methods were also devised, so that the most relevant results can be highlighted automatically, modifying the threshold or the metric used.

Secondly, once tool development is complete, the tool should be validated by comparing the results it yields with those obtained by a group of experts. To do so, appropriate evaluation devices should be developed.

Finally, usability testing strategies will have to be developed for the software tool. As mentioned before, a significant requirement for this tool is that non-expert users can effectively use it; additionally, it should also simplify and clearly explain the results obtained. Thus, it will also have to be evaluated considering these aspects in addition to technical assessments based on expert judgment.

References

1. Alibrahim, H., Ludwig, S.A.: Hyperparameter optimization: comparing genetic algorithm against grid search and bayesian optimization. In: 2021 IEEE Congress on Evolutionary Computation (CEC), pp. 1551–1559. IEEE (2021)

2. Bianco, S., Lanzarini, L.C., Zangara., M.A.: Evolución temática de publicaciones en español. una estrategia posible para el diseño de situaciones didácticas. In: XVI Congreso de Tecnología en Educación y Educación en Tecnología (TEyET 2021), RedUNCI (2021)
3. Bianco, S., Lanzarini, L.C., Zangara, M.A.: Thematic evolution of scientific publications in spanish. In: XXVII Congreso Argentino de Ciencias de la Computación (CACIC)(Modalidad virtual, 4 al 8 de octubre de 2021) (2021)
4. Callon, M., Courtial, J.-P., Laville, F.: Co-word analysis as a tool for describing the network of interactions between basic and technological research: the case of polymer chemsitry. Scientometrics **22**, 155–205 (1991)
5. Callon, M., Courtial, J.-P., Turner, W.A., Bauin, S.: From translations to problematic networks: an introduction to co-word analysis. Soc. Sci. Inf. **22**(2), 191–235 (1983)
6. Cobo, M., López-Herrera, A., Herrera-Viedma, E., Herrera, F.: An approach for detecting, quantifying, and visualizing the evolution of a research field: a practical application to the fuzzy sets theory field. J. Inf. **5**(1), 146–166 (2011)
7. Coulter, N., Monarch, I., Konda, S.: Software engineering as seen through its research literature: a study in co-word analysis. J. Am. Soc. Inf. Sci. **49**(13), 1206–1223 (1998)
8. P. K. Project. Open journal system (2001). http://pkp.sfu.ca/ojs
9. Ramos, J., et al.: Using tf-idf to determine word relevance in document queries. In: Proceedings of the First Instructional Conference on Machine Learning, vol. 242, pp. 29–48. Citeseer (2003)
10. Robertson, S.: Understanding inverse document frequency: on theoretical arguments for idf. J. Documentation (2004)
11. Sternitzke, C., Bergmann, I.: Similarity measures for document mapping: a comparative study on the level of an individual scientist. Scientometrics **78**, 113–130 (2009)

Higher Education and Virtuality from an Inclusion Approach

Javier Díaz⬤, Ivana Harari⬤, Ana Paola Amadeo⁽✉⁾⬤, Alejandra Schiavoni⬤, Soledad Gómez⬤, and Alejandra Osorio⬤

LINTI – Research Laboratory in New Computer Technologies,
School of Information Technology, National University of La Plata, La Plata, Argentina
{jdiaz,aosorio}@unlp.edu.ar, {iharari,ales}@info.unlp.edu.ar,
pamadeo@linti.unlp.edu.ar, sgomez@cespi.unlp.edu.ar

Abstract. The pandemic that started at the beginning of the year 2020 and the isolation situation virtualized face-to-face educational proposals very quickly. The distance modality confronted us with the review of teaching practices including the planning, evaluation and monitoring of students and an exhaustive evaluation of the tools to be used. In this period, videoconferencing tools emerged as the only resource that allowed synchronous meetings and collaborative interaction. In addition, the use of educational platforms and accessible resources was necessary to ensure that the student can interact with the teacher and the study material. In this article, the accessibility analysis of videoconferencing systems used by students with disabilities in the Computer Science School subjects of the National University of La Plata is presented. The functionalities provided for disability, their compliance with international accessibility standards were analyzed and the tools were also manually tested in different interaction scenarios. Besides, different tools that allow validating the platforms used and creating accessible material for students are mentioned. A set of recommendations is also proposed as a contribution to improve the teaching strategies with technologies, considering the different situations of students with disability.

Keywords: Accessibility · Video conferencing systems · Educational digital resources · Pandemic education

1 Introduction

The isolation situation caused by the pandemic made it not only essential but urgent at the beginning of the 2020 school year to adapt teaching strategies at all educational levels. The distance modality, which we experienced for the first time and without enough prior preparation, confronted us with the revision of our teaching practices, including the course planning, the evaluations and the students monitoring.

At the same time, tools such as web conferencing, that is a bidirectional and synchronous tool, emerged as the only resource that allowed us to have a simultaneous meeting and interact collaboratively. The importance acquired of these tools during periods of

P. Pesado and G. Gil (Eds.): CACIC 2021, CCIS 1584, pp. 78–91, 2022.
https://doi.org/10.1007/978-3-031-05903-2_6

isolation was related to a pedagogical aspect, the consolidation of links between people, the virtual community, and a social aspect, as it allowed the inclusion of participants who are geographically dispersed. A few years ago, web conferencing systems were quite limited, but today there exist interesting videoconferencing systems that offer a variety of functionalities, constituting a tool with significant pedagogical potential.

Although videoconferencing tools, in many cases powerful, allow us a certain approach, this is not comparable to face-to-face. The virtual environment, the intermediary software, the communication problems, the interference, the noise, make the virtuality present and the distance is really felt. In the case of people with disabilities, videoconferencing tools limitations (communication problems, interference and noise) are further deepened, hindering the educational practices developments that promote training and learning processes. Not being able to connect, to access the full material, to hear what the teacher is explaining, or to see the slides that the teacher is sharing via videoconference, further increase the distance. This gap generates a distance that is difficult to overcome if it is not through the implementation of specific strategies that promote consensual alternatives with the 1,500 students with disabilities throughout the university. The inclusion of these dynamics is configured as bridges that the teachers must enable, that must be built together with the person with disabilities.

The topic of web accessibility represents a very relevant line of research, which has been worked on in the Computer Science School since 2002, and which was incorporated into the study plan of the Bachelor of Computer Science and Bachelor of Systems careers. It is important to note that its approach was institutionalized through the creation of an Accessibility Department since 2010, which has close contact with students with disabilities.

In this article, the open platforms used in the Computer Science School are presented through which the totally virtual courses were implemented. Based on its massive use, the Moodle learning management system that complies with several accessibility guidelines, according to international standards, is analyzed and described. The accessibility analysis of the videoconferencing systems used by students with disabilities in the subjects is presented, from an approach that includes: 'Inclusion in education consists of ensuring that each student feels valued and respected and can enjoy a clear sense of belonging. However, many obstacles stand in the way of that ideal. Discrimination, stereotypes, and alienation exclude many. These exclusion mechanisms are essentially the same regardless of gender, location, wealth, disability, ethnicity, language, migration, displacement, sexual orientation, imprisonment, religion, and other beliefs and attitudes.' [1] From this perspective, it is necessary to consider virtual learning and teaching platforms, digital resources, reference web pages, ICTs used by people with disabilities, among other issues. In the article *Contributions to think education in a pandemic from accessibility* presented in CACIC 2021 [2], the complexity of the subject is recognized, analyzing different products, which cover the points mentioned previously. In addition to the Moodle virtual platform, videoconferencing systems are analyzed and compared: Zoom, Big Blue Button and Webex. For this, a series of tests and checks were carried out, such as, for example, the functionalities provided for disabilities, their compliance and compliance with international accessibility standards. All tools were also manually tested in different interaction scenarios. This article extends the analysis carried out incorporating the students with disabilities interviews, in which they expressed their personal experience with the use of each of the platforms. In addition, a set of tools are mentioned that allow validating the educational platforms used and creating accessible material in different formats for students.

Important recommendations to take into account regarding accessibility of the video-conferencing systems analyzed and of the teaching strategies considering the students with disabilities particular conditions are also reflected.

2 Virtual Education and Accessibility

Thinking about accessibility issues in the context of remote education or emergency education poses a challenge. The speed with which the face-to-face educational proposals were virtualized in the context of isolation, homogenized the pedagogical strategies, which in matters of accessibility deserve a treatment from diversity. Many factors come together when academic activities are transferred to a totally virtual plane that affects the teaching and learning process, changing its entire context. What used to be manifested in a classroom, now combines: the conditions of each subject, be it a teacher, a student or an assistant therapist or interpreter in the case of students with disabilities, who intervenes in the educational process. The technological environment, the specific characteristics of each participant who takes part in the training process, the software chosen for academic activities, whether it is for a repository of educational materials and resources, for communication, for classes in videoconferences, for the evaluation, it configures 'a virtual classroom' unique, specific and particular for each one. This affects in different ways, their level of perception, understanding, communication, acquisition of knowledge and subsequent application.

Understanding that the group of people with disabilities is by no means homogeneous, progress is made towards the idea of realizing true educational inclusion for all students from the perspective of rights and based on particularity and individual differences. We assume the challenge of working under the social paradigm approach that becomes the basis of some of the pedagogical interventions that we carry out and on which we investigate. It makes us think the work with disabilities as 'on educational integration, and concepts such as educational needs derived from disabilities take relevance, in the intention of marking that the person's disability is more linked to barriers to learning and participation (Booth and Ainscow 2002) produced in the environment than to the deficiency itself' [3].

In regulatory terms, Law 24,521 on Higher Education of the Argentine Nation, which universities, university institutes and higher education institutes must respect, stipulates that educational inclusion policies must be in place, taking measures to give equal opportunities and possibilities to people with disabilities, and that they can access the system without discrimination of any kind [4]. In turn, the spirit of the Statute of the National University of La Plata [5], which from its preamble defines it as a public and free higher education institution, which must be offered open and inclusive for the entire society, establishing policies that tend to facilitate the permanence of the most vulnerable sectors of society.

At the level of digital accessibility, Argentina has had a legislative framework since 2010 when Law 26,653 on accessibility of information on web pages was unanimously approved [6]. It stipulates that national government sites must respect the accessibility standards of the W3C, specifically levels A and AA of WCAG 2.0 [7]. As the UNLP is a national institution, the sites belonging to the domain unlp.edu.ar are included in the scope of this regulation [8].

The challenge of integrating, associated with the responsibility of promoting educational spaces based on the principles of equality and inclusion, transcends the barriers of face-to-face, virtual, synchronous, and asynchronous modalities as they are part of the ethical responsibilities to which we, as teachers, have committed ourselves. Understanding and rethinking teaching practices from complexity and in pursuit of equality and inclusion will allow us to think of the University as 'The school of inclusive design thus becomes part of a philosophy of life that recognizes diversity as a value and as a source of enrichment, whose success is reflected in the achievements of each one of the students and in the development and well-being of the community. This type of education for inclusion -conceived as education for life- should be accessible especially to the most vulnerable and marginalized groups, highlighting the importance of this continuous and gradual process starting early and being able to be articulated throughout the different levels of education' [3].

3   Accessibility Features of the Moodle Learning Management System

Virtual platforms or LMS are today fundamental tools for learning and teaching through the Internet. In the current market, more and more companies are adding new platforms to their offered products, with Moodle [9] as an open-source product and Blackboard [10] being the oldest and best positioned in terms of its use and versatility of the offered solution. Although it is a reality that each solution has its pros and cons, when choosing it will depend on the resources identified as priority, license costs, and infrastructure services offered, among others.

Moodle is described below, being a de facto standard in open source LMS, in addition to being the tool that has been used since 2003 in the Computer Science School, in a wide variety of courses, in addition to being among the first in global trends as an e-learning platform and be proposed as a virtual platform for online courses by the SIU (University Information System) consortium of the National Universities of Argentina [11]. The Moodle core developers are committed to using Web standards, the Web Content Accessibility Guidelines WCAG 2.1 level AA [12], Authoring Tool Accessibility Guidelines ATAG 2.0 [13] for content generation, Accessible Rich Internet Applications ARIA 1.1 [14] for rich content to create engaging and interactive presentations as well as US section 508 regulations. It is also important to note that the platform has many modules, which users enable and disable, use custom themes and it is possible to make different configurations according to the needs of the institution and the knowledge of the administrator. This implies that it is impossible to ensure 100% accessibility on the platform. However, the following sentence illustrates Moodle's philosophy 'Accessibility is not a state, it is a process of continuous improvement in response to our users and the greater technical environment' [15]. The ATAG Report Tool [16] provided by the

W3C to assist in the construction of reports that allow evaluating the accessibility of a content generation tool is interesting.

Moodle's Atto text editor [17] allows you to generate accessible content. It offers configuration options through the Administration panel. From the teacher profile or course editor, this editor offers accessibility validations according to the HTML5 standard and the WCAG from the toolbar, such as the accessibility check button and the screen reader.

Moodle also has plugins that contribute to accessibility such as the Accessibility block [18], which allows you to customize the view by increasing the font, changing the contrast, and integrates the ATbar, an open source, cross-browser product, developed by Southampton University ECS [19]. It also allows the administrator to customize the user interface through accessible themes [20], being Moove and Adaptable the most downloaded. As can be seen, not only the learning management platform must be accessible, but also the content that is generated and the analysis of the complementary tools for virtual classes, such as synchronous sessions using videoconferencing tools. The following section presents an accessibility analysis of the videoconferencing tools used during this period.

4   Analysis of Videoconferencing Tools According to Disability

As mentioned above, due to the health context, it gave rise to mobility restrictions and preventive social confinement, which had to be implemented in a mandatory and rigorous manner in most countries, and Argentina was no exception. The use of digital technological devices and media, specifically the Web, became a demanding and mandatory medium, mainly to continue in the teaching and learning process.

Face-to-face classes became synchronous meetings through videoconference systems, consultations, forums, evaluation instances, remote examination systems, with cameras and microphones on, trying to ensure that the educational process is not interrupted and can be established. In the case of people with disabilities, it is not only enough that the classes, materials, and other educational resources are available on the Internet, but they must also be accessible.

For this reason, this article tries to analyze the level of accessibility of the digital tools used to carry out remote synchronous classes, such as videoconferencing systems, from the perspective of higher education students with different disabilities.

Specifically, the tools Zoom v5.4.6 [21], Big Blue Button 2.2 [22] and Webex 39.3.0 [23] were analyzed, which were used in the subjects where there were students with disabilities from the School. This section details the analysis performed with automatic tools such as Wave [24] and the manual validations were carried out using the NVDA screen reader [25], the Web Developer extension [26], the Colour Contrast Checker [27] the Windows virtual keyboard and magnifying glass, Insights for Web [28]. Below is a comparative table between the different tools, regarding the degree of compliance in terms of:

- Automatic accessibility validators: tools have been validated with the accessibility validators. It is possible to meet an adequate level on all screens or have difficulties in specific sections. For example, in WebEx, the validator detects some accessibility problems such as contrast problems and some icons without alternative texts, it also presents some problems in the structure of headings and some sectors lack them.

- Screen readers: The tool was used with the screen reader enabled. Zoom presented no major difficulties but in BBB, the reader does not read the "join as listener" window or that the audio device is being activated. For its part, Webex does not read the copy invitation icons, nor does it read the update calendar, and it reads the internal name of the component, such as button 501139, making the interaction very confusing. Also, the reader does not read the Help icon. And some phrases appear in English. This feature is particularly relevant for people who are blind or have low vision.

- Contrast: The appropriate level of contrast has been evaluated in different scenarios such as when using small fonts or the magnifying glass. The 3 tools presented problems in certain sectors where very small letters, icons and buttons whose click area is less than recommended are used. This is particularly significant for people with low vision.

- Responsive design: the tool automatic adaptation to different devices is relevant to operate the software properly, regardless of the device used. In this sense, BBB allows access to functions that are not displayed in the window frame. While Webex also presents some problems in this regard.

- Subtitling: the possibilities of the tools for the adequate management of subtitles were analyzed, such as the possibility of having online subtitles or a special role during the sessions or the possibility of recording the classes for later subtitling. This feature is particularly relevant for deaf people.

- Keyboard navigation: Only the keyboard was used to interact with the tool. Checked for proper tab order and shortcuts to main functions. While the 3 tools did not present any major difficulties, WebEx did present some difficulties in configuring the audio that does not have an output option and in handling the level up/down sliders. It also presents some pop-up menus, which are displayed by pressing the dotted icon that is cyclical if it is operated from the keyboard with the TAB key. This is correct, but you should have a cross to get out of there. This feature is particularly relevant for people with motor disabilities. Figure 1 shows an example of a navigation scheme using the keyboard in BBB.

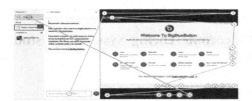

Fig. 1. Scheme of navigation using keyboard in BBB

- Browser with JavaScript and style sheets disabled: the only case that applied this validation was BBB, whereby disabling style sheets and JavaScript it is possible to interact properly. Particularly relevant for browsing using different devices and old versions of browsers.

Table 1 presents a summary of the analysis carried out for the different videoconferencing tools mentioned.

Table 1. Summary of the accessibility features of the VC tools analyzed

	Zoom	WebEx	BBB
Screen reader	Apply	Partially apply	Partially apply
Responsive design	Apply	Present problems	Present problems
Accessibility validator	Apply	Partially apply	Partially apply
Disable styles and JavaScript	Not apply	Apply	Not apply
Online subtitling	Not apply	Not apply	Not apply
Role for manual subtitling	Apply	Apply	Apply
Transcript of recorded meetings	Apply	Apply	Apply
Session recordings	Apply	Apply	Apply
Keyboard navigation, tab order	Apply	Apply	Apply
Keyboard navigation between the most important options	Apply	Apply	Partially apply
Contrast	Partially apply	Partially apply	Partially apply
Lens	Apply	Apply	Apply

Taking only the points that apply to the 3 tools, it can be seen that Zoom complies with 81.81% of the aspects evaluated, while BBB complies with 63.63% and WebEx with 45.45%. For its part, Zoom partially complies with 9.09% of the aspects evaluated, while BBB partially complies with 27.27% and WebEx with 36.36% of the characteristics analyzed. Finally, Zoom does not meet 9.09%, BBB and WebEx do not meet 18.18% of the mentioned features.

5   Analysis of Videoconferencing Systems by Students with Disabilities

In this situation of compulsory virtuality, the School has had close contact with students with disabilities, highlighting the problems encountered in their exclusively distance learning process. A virtual discussion-type meeting was held, in which 15 students participated. Four of them were deaf, three blind, two with dyslexia, three with Asperger syndrome and one with cochlear implant. The videoconferencing systems used in the courses were analyzed in the meeting and how was the experience they had with them. Initially, they were consulted in which videoconferencing system they wanted to hold the meeting, so as to have an idea of their preference in the first instance. 14 of the 15 indicated Zoom, and the remaining person replied that he did not care.

During the conversation, the following trigger questions were asked: what do they think of virtuality? How is the experience in virtual classes? In which videoconferencing systems were they held? Did the digital device used affect participation during class? What problems did they find? During the meeting, the following interventions were recorded: the simplest was Zoom, it is the lightest (all); the Zoom system can be used without problems from the mobile (blind students); Webex requires a lot of resources, does not work well on Linux, it is necessary to install more resources for it (the student with dyslexia); it is important that the interpreter and the teacher are fixed on the screen and that is possible in Zoom (deaf students); Webex usually stops, and takes a long time to load, when connecting again the class is very advanced (deaf students); Zoom leaves the files in.mp4 format, they can be uploaded to YouTube and then subtitled (hard of hearing and deaf people); BBB is heavy from the cell phone since you also must enter Moodle as well (blind people); in BBB the reader does not read if you enter with a microphone or audio (blind people); the format of BBB's recorded videoconferences is not versatile; it does not generate an easily manipulated file (most of them); the need to have the slides of the classes in advance (everyone); the teacher should describe in detail what is on the slides to be able to follow the class well (blind students); the teacher should speak and articulate the words well to allow them to read their lips (hard of hearing); consultation classes should be held in virtual meetings. Forums are not enough. In them, the questions and answers are scattered and disordered, and in many cases the answers must be followed in the conversation threads that are generated (all of them).

6 Practices and Tools to Create Accessible Courses

Designing an accessible course requires thinking of a course that is accessible to any type of student, that is, they can receive information, interact with the study material, with the teacher and their peers and demonstrate their learning beyond their age, gender, race, language, disability, etc. The universal learning design framework [29] is a guide of recommendations to generate a flexible learning environment that aims to improve perception, expression and understanding. In this sense, the digital learning-teaching platforms, the study material, the communication channels and all the tools must be accessible. Providing multiple options does not mean lowering expectations but creating an inclusive space ensuring that learning takes place for all participants.

Regarding the digital resources of the course, it is essential to ensure that the readings, the videos, the synchronous and asynchronous communication tools are accessible. Below is a series of tools that help the teacher to develop and validate the material and tools used.

- Documents: simple textual content, which facilitates searches, and the creation of indexes are usually accessible. Furthermore, according to the WCAG 2.1 level AAA navigable principle, it is recommended to use the heading hierarchies provided by different text editors, such as MS Word or GoogleDocs or OpenOffice. Use simple fonts, such as serifs size 12 or larger [30]. Check the contrast of the letters with the background and use double spacing. Only hyperlinks should be underlined. In the document properties, set the language of the document. For long documents, it is suggested to include an index or table of contents available in text editors. The same recommendations apply to presentations like PowerPoint or GoogleSlides. In Microsoft products, use the Accessibility Checker [31] to detect and correct accessibility problems.

- Accessible PDFs: Version 1.7 of the Portable Document Format became an open standard: ISO 32000–1:2008 [32], making it a standard format for creating documents. The ISO 14289–1:2014 PDF/UA standard [33] describes the required and prohibited components, and the conditions for their inclusion in a PDF file to make it accessible by the largest number of users. The WCAG 2.0 standard defines 23 techniques to comply with in PDF documents, making use of the features provided by PDF files. The core of this support is basically to create tagged PDFs and the ability to determine the logical order of the content in a document, regardless of its appearance. This logical order is built by incorporating structural and semantic information through tags, like HTML or XML. These tags allow you to define a tree-like hierarchy that makes it easy to navigate. ISO 32000 defines in its section 14.8 the rules for tagged PDFs and a set of standard tags. PDF files can be created from accessible documents, the considerations mentioned in the previous section also apply to this type of document. Scanned documents need to be OC Red to recognize text in scanned images. Acrobat Reader Pro [34] allows you to create accessible documents, detect and correct accessibility problems. Pave [35] is an online service for PDF editors and it is also an open-source product, which makes it easy to detect and correct errors. Tingtun Checker [36] is an online validator for PDF files and web pages, in English and Polish. It validates PDF files according to WCAG 2.0 standards.
- Links: WCAG 2.1 makes a series of recommendations related to the Operable objective regarding links. At level A, links must be texts that connect to other documents with a clear and concrete purpose in itself or through its context. Avoid the use of the URL or the "click here", replacing it with the name or description of the destination. Warn the user if the memory pointed to by the link requires an additional application to be opened. At the AAA level, it is recommended to use specific labels to indicate the purpose, provide alternative texts for images or graphics that are links, among others. It is important to detect and correct broken links or those that do not lead anywhere. There are numerous tools to perform this task, such as Broken Link Checker [37], a Chrome plugin, SiteChecker that provides a free version to perform online validations [38], LinkChecker an open-source product for detecting broken links [39], Link Crawler [40] is another open-source product.
- Tables: Tables must include descriptive labels, providing information about the rows and columns. The first row usually identifies the header. In addition, cell grouping should be avoided, and its reading checked using the Tab key. Include a title according to the content of the table, alternative texts and the 'repeat table header' property. Check the contrast between the letter, the background, and the edges.
- Images and graphics: WCAG 2.1 make a series of recommendations related to the Perceivable principle. According to the recommendation 1.1.1 Non-text Content, level A, 1.4.5 level AA and 1.4.9 level AAA, images and graphics must be accompanied by a descriptive alternative text and a long description that represents the image content. It is advisable to avoid images with textual content. These texts are read by screen readers instead of images or graphics. Include references to images or graphics within

the text. Avoid flashes and in case of including animations or 'carousels', include commands for their control by the user, according to the Operable principle of the standard. If the content is a Web page, it is possible to use the Chrome Web Developer extension [26] to display the page without images and verify that there is no loss of information.

- Videos: WCAG 2.1 includes a series of recommendations for audiovisual content [41]. The accessibility of a page depends not only on the HTML code but also on the elements that are included in it, so the accessibility of a page will depend on the accessibility of its entirety. If the video is going to be distributed in pre-recorded format, the accessibility requirements established by the WCAG 2.1 standard regarding level A, Closed Caption (CC) subtitling and transcription in text or audio description format are required. As for AA level, CC subtitling, transcription in text format and audio description are required. Level AAA requires extended audio description, sign language. If the video is broadcast live, the material must comply with level AA, it must be subtitled. Interpretation in sign language and the rest is not required by the standard. It is advisable to publish them before the class. Students will be able to access at any time and place, with different ways to access the content, with text being the main alternative means. There are multiple tools to record classes, both open-source desktop tools such as OBS Studio [42], as well as browser plug-ins, such as the Awesome screenshot extension [43] or Loom [44]. The Center for Teaching Innovation at Cornell University [45] gives a set of recommendations, which include tips for before, during, and after the recording process of classes, such as checking the light, practicing before recording, videos of 20 min maximum duration, among others. The inclusion of a sign interpreter is regulated by UNE139804:2007- Requirements for the use of Spanish sign language in computer networks. Among other points, it gives recommendations that the image of the interpreter in sign language be clear with good lighting; ensure that the upper torso is well observed, where the expressions of the face such as the gestures of the hands and arms are well exposed; perform a good montage and synchronization between the original video and the video of the performance, among other aspects.
- Audio and subtitles: As mentioned in the previous item, the standard determines the aspects to be considered to comply with the Perceptible principle of the WCAG 2.1 standard. It is very important to provide subtitles as they compensate for noisy environments or where sound is not allowed, they provide a solution when the audio has poor quality, it allows you to quickly navigate through a video by reading the subtitles by scrolling the video control. Although there are different types of subtitles, for accessibility reasons, Closed Caption type subtitles are essential, where the user can activate or deactivate depending on the needs of it. The most common subtitle formats are: Subrip, SMIL and WebVTT. Among the tools to generate subtitles we can mention Express Scribe [46], a subtitling tool for Windows and Mac, in Spanish; it offers a free reduced version and a paid professional version; CapScribe [47], a free tool for Mac and Windows, allows you to subtitle videos in different formats, version 2 is in the testing stage; Gnome subtitles [48], free subtitling tool for Linux; Gaupol [49], a free tool for Windows that allows subtitling, among other features. Subtitle Horse [50], free online subtitling tool; OpenCaps [51]: free online subtitling tool. Allows you to export subtitles in various formats: DFXP, DVD STL, MicroDVD, QT Text, SAMI, SubRip

and SubViewer; and finally Subtitle Workshop [52], a free subtitling tool for Windows that exports subtitles in multiple formats. For its part, YouTube has voice recognition tools to automatically generate subtitles. Audio description is another important point to consider, which transforms visual information into auditory information, taking advantage of blank spaces and through voiceover. It describes actions, scenarios, and locations, reading texts that are presented on the screen, among others. The standard that regulates audio descriptions is UNE 153020.

- Automatic accessibility validators: accessibility validators are very useful and essential tools for developers and for content generators, both for Web pages and other digital resources that we mentioned previously. These tools allow you to detect accessibility problems, generally according to the W3C WCAG 2.1 standards. For the validation of Web pages, we can mention Wave [24], a suite of tools to validate and teach about the accessibility of Web pages developed by WebAIM of Utah State University, includes extensions for Chrome and Firefox, English language; Tawdis [53], online Web page validator, which provides very detailed reports, in Spanish, Portuguese and English. The W3C provides a list of tools to check the accessibility of different resources according to different standards [54].

- Screen readers: screen readers are also often used to perform manual accessibility validations; however, their learning curve is steep and they tend to throw false positives more related to the use of the tool by the user than to real problems. Among the most used we can mention JAWS for Windows [55], NVDA, an open-source product previously mentioned, Orca for Linux [56], MS Narrator integrated into MS Windows [57], Talkback for Android devices [58], Voiceover and Siri [59] for iOS devices.

7   General Recommendations

As a result of the analysis carried out and recovering an educational approach that includes inclusion of people with disabilities, a series of recommendations can be made for teachers who use video conferencing systems such as Zoom, Webex and Big Blue Button (BBB).

It is important to consider the conditions of students with disabilities, if they have the technological resources that are required when using virtual distance strategies. In this sense, agreeing on communication ways and strategies is an important factor to adapt the classes and the educational materials. In the case of students with disabilities, it is essential to bring them personalized meetings with few people, direct communication channels or more time for resolving the problems and exercises. For example, deaf students can read lips, or require the presence of the interpreter. The same goes for people with dyslexia and dysplasia, and those with motor problems that make typing difficult.

In terms of teaching strategies, adapting times, strategies and materials allows for better organization, it is important to take into account the time it may take to adapt materials so that they are suitable for students with disabilities. Therefore, the demands on tasks delivery times must be extended. Along these lines, it is recommended that when sharing PDF documents, videos, slides, or other educational material, it be accessible and, if possible, previously delivered to the student. It is very useful to establish together with students with disabilities, the required formats for the materials that are published and for the productions that are requested of them.

Understanding the context of interaction of students with disabilities enables us to design strategies and select tools from an accessible perspective. For example, blind students need to handle everything using the keyboard and that the application used supports screen readers such as Jaws, NVDA or cell phone readers. In the case of deaf students, they require subtitling or sign language interpretation, so the choice of video conferencing or video call applications that automatically subtitle or at least through a subtitle role is recommended.

Regarding the pedagogical decisions on the technique, it is recommended to verify that digital resource access is accessible, the invitation to the meeting must be granted in several ways. WhatsApp, email, the Moodle platform present accessible features. And it is important that the videoconferencing application itself is accessible.

Thinking about and analyzing the functionalities and accessibility services offered by videoconferencing applications are actions that will allow us to foresee that, for example, for deaf people, what teachers are speaking is required to be subtitled. Or they will allow us to adapt the recordings that are made of the meetings. It is important that all meetings held by the chair be recorded so that the student with disabilities has access and review after the meeting, they must incorporate subtitles while it is necessary to analyze the visual content that was not notified by the teacher and communicate it to the blind person.

Also, in relation to the materials, it is essential to make accessible what is shared on the screen or notified it. Teachers often share slides or documents on the screen. These must be accessible and, if possible, have been given to the students previously. There are some limitations that condition the possibilities of work, for example, screen readers do not read the material shared by videoconferencing systems, since the audio is held by the teacher who is giving the class. It is recommended that the teacher avoid location words such as: 'Here you are…', 'You should pay attention to this…', 'How do you see here…', 'This and what we did there…'.

Lastly, it's important to verify that your video conferencing app's notifications work on screen readers. These recommendations are contributions to deepen the approach of teaching strategies with technologies, in recognition of the multiple realities presented by students with disabilities.

8 Conclusions

This article made an analysis from the perspective of the accessibility of digital educational resources and platforms that were used and adopted in the pandemic period in the Computer Science School of the National University of La Plata. It determined other forms of communication, participation, planning, problem solving and conflict resolution in the teaching process, of which several can be sustained in the post-pandemic period. The accessibility of the Moodle virtual platform was analyzed and the videoconferencing systems that are being used to supplant face-to-face classes. A set of tools for assessing accessibility and for creating accessible content were also mentioned. It is important to note that considering a complex approach to education, the tools are not the only conditions to be considered in the teaching process. In this sense, the particularity of each student from their perspective in their virtual environment, the teacher's training regarding technical issues and the use of technological devices, the software applications used in their teaching process, they are as relevant as the specificity of the content to be taught and the approach adopted by each teacher for work in the classroom. These aspects require adaptation channels, consensual and well-designed levels of stimulation and containment, considering the specificity of the student, the context and the resources used.

Regarding the specific case of videoconferencing systems, they constitute a tool with significant pedagogical potential, but it is necessary to analyze them with respect to their accessibility services to understand the impact that they can generate when they are used by people with disabilities. They must be selected and used in such a way that all students can enjoy this resource in the most equitable way possible. Accessibility characteristics must be considered if the aim is to guarantee effectiveness and efficiency in synchronous meetings for all students. For this reason, in this article a contribution is made on their accessibility level, considering their services, if they comply with the conformity of the WCAG, if they support different interaction scenarios and most importantly, how is the impact in students with disabilities who remotely, from a distance and without assistance had and have to access classes and understand them.

In addition, a series of recommendations and tools focused on the production of digital and accessible teaching materials are proposed, which recognize the conditions of students with disabilities and promote the readjustment of teaching strategies to achieve true inclusion of all students. The conditions of access and use of these and other tools constitute a challenge to continue exploring (in economic, licenses, installation and learning aspects) to think about inclusion from an integration approach in recognition of diversity.

References

1. UNESCO. 2020. Summary of Monitoring report of World Education 2020: Education and Inclusion: without exceptions. https://bit.ly/3I4B1Vk
2. Díaz, J., Harari, I., Schiavoni, A., Amadeo, A., Gómez, S., Osorio, A.: Aporte para pensar la educación en pandemia desde la accesibilidad. In: Proceedings of Congreso Argentino en Ciencias de la Computación CACIC 2021. 1a ed. (2021). ISBN 978-987-633-574-4
3. Sonia L.B., et al.: Infancia, discapacidad y educación inclusiva: investigaciones sobre perspectivas y experiencias 1a ed. - La Plata: Universidad Nacional de La Plata. Facultad de Psicología 2019. http://sedici.unlp.edu.ar/handle/10915/83689
4. Ministerio de Justicia y Derechos Humanos. Ley de Educación Superior Ley 24. 521. http://servicios.infoleg.gob.ar/infolegInternet/anexos/25000-29999/25394/texact.htm
5. Honorable Asamblea Universitaria: Estatuto de la Universidad Nacional de La Plata (2009). https://unlp.edu.ar/frontend/media/20/120/722e7f1b616ac158e02d148aaeb762aa.pdf
6. Congreso Nacional de la República Argentina; Ley 26653 Accesibilidad de la información en las páginas web. https://bit.ly/3t3NW5C
7. Web Content Accessibility Guidelines (WCAG) 2.0. https://www.w3.org/TR/WCAG20/
8. Disposición 6/19 de la (ONTI). https://bit.ly/3KwQoHE
9. Moodle. https://moodle.org/
10. Blackboard. https://www.blackboard.com/teaching-learning/learning-management

11. Interface SIU-Guaraní-Moodle. https://bit.ly/3J8mA3P
12. Web Content Accessibility Guidelines (WCAG) 2.1. https://www.w3.org/TR/WCAG21/
13. Authoring Tool Accessibility Guidelines (ATAG) 2.0. https://www.w3.org/TR/ATAG20/
14. Accessible Rich Internet Applications (WAI-ARIA). https://www.w3.org/TR/wai-aria-1.1/
15. Accesibilidad en Moodle. https://docs.moodle.org/all/es/Accesibilidad
16. ATAG Report Tool. https://www.w3.org/WAI/atag/report-tool/
17. Editor Atto. https://docs.moodle.org/all/es/Editor_Atto
18. Block accessibility, Moodle plugin. https://moodle.org/plugins/block_accessibility
19. ATbar. https://www.atbar.org/
20. The Best Free Themes for your Moodle-based Learning Environment EVER (New Decade Update). https://bit.ly/3MN6ejA
21. Zoom: Accessibility Features: zoom.us/accessibility
22. Big Blue Button Accessibility. https://bit.ly/3KTUPN9
23. Webex Meetings Accesibilidad. https://bit.ly/3t6dvTL
24. WAVE Web AccessibilityEvaluation Tool. https://wave.webaim.org/
25. NVDA en Español. https://nvda.es/
26. Web Developer. https://bit.ly/3I3fexm
27. Colour Contrast Checker. https://bit.ly/3t3Ncxm
28. Accessibility Insights for Web. https://accessibilityinsights.io/docs/en/web/overview/
29. Universal Design for learning. https://bit.ly/3KFEDi3
30. Freyhoff, G., Hess, G., Kerr, L., Menzel, E., Tronbacke, B., Van Der Veken, K.: El Camino Más Fácil. Directrices Europeas para Generar Información de Fácil Lectura.AsociaciónEuropea ILSMH, Ed.Portugal (1998). https://bit.ly/3i1KeTM
31. Improve accessibility with the Accessibility Checker. https://bit.ly/3tU7zMR
32. ISO 32000-1:2008. https://www.iso.org/standard/51502.html
33. ISO 14289-1:2014. https://www.iso.org/standard/64599.html
34. Create and check accessibility of PDF (Acrobat Pro). https://adobe.ly/3hYJGxN
35. PAVE. https://pave-pdf.org/
36. Tingtun Checker. https://checkers.eiii.eu/en/pdfcheck/
37. Broken Link Checker. https://error404.atomseo.com/
38. SiteChecker. https://sitechecker.pro/website-crawler/
39. LinkChecker. https://github.com/linkchecker/linkchecker
40. LinksCrawler. https://github.com/rakeshmane/Links-Crawler
41. How to Meet WCAG. https://www.w3.org/WAI/WCAG21/quickref/#time-based-media
42. OBS Studio. https://obsproject.com/
43. Awesome screenshot. https://www.awesomescreenshot.com/
44. Loom. https://bit.ly/3CAQuex
45. Pre-recorded Lecture Tips. https://bit.ly/3CAQE5D
46. Express Scribe Transcription Software. http://www.nch.com.au/scribe/
47. Inclusive media & design. Web accessibility for all. https://bit.ly/3pUbs2Y
48. Gnome Subtitles. http://gnomesubtitles.org/
49. Gaupol. Editor for text-based subtitles. https://otsaloma.io/gaupol/
50. Subtitle Horse. Subtitle and captions editor. http://subtitle-horse.com/
51. OpenCaps. Caption and subtitle editor. https://opencaps.idrc.ocadu.ca/site/
52. Subtitle Workshop. https://subtitle-workshop.en.lo4d.com/windows
53. TAWdis. Web accessibility test. https://www.tawdis.net/#
54. Web Accessibility Evaluation Tools List. https://bit.ly/3pTjclV
55. Dolphin Supernova. Magnifier, & screen reader. https://bit.ly/3w5qiI1
56. Orca screen reader. https://help.gnome.org/users/orca/stable/index.html.es
57. Microsoft Windows Narrator. https://bit.ly/3i5GwIA
58. Android Accessibility Suite. https://bit.ly/37bL2Dh
59. Apple accessibility. https://www.apple.com/accessibility/

Graphic Computation, Images and Visualization

Automatic Extraction of Heat Maps and Goal Instances of a Basketball Game Using Video Processing

Gerónimo Eberle[(✉)], Jimena Bourlot, César Martínez,
and Enrique M. Albornoz

Research Institute for Signals, Systems and Computational Intelligence,
sinc(i) UNL-CONICET, Ciudad Universitaria, Ruta Nac. No. 168, Km 472.4, (3000),
Santa Fe, USA
contacto@geronimoeberle.com.ar,
{cmartinez,emalbornoz}@sinc.unl.edu.ar

Abstract. The use of image processing techniques and computer vision to obtain teams statistics in different sports, currently represents a new source of information very useful for season preparedness. In this work, we propose different methods to show the players position for a specific time, to perform the point counting, and to extract clips of goal situations in the match. In order to accomplish this, we use a combination of transfer learning using a pre-trained deep neural network with a database of basketball game excerpts, and video processing techniques. As a proof of concept, the method was applied to a basketball game of local teams, showing the feasibility of the proposed approach.

Keywords: Game statistics · Heat maps · Digital image processing · Deep neural network · Transfer learning

1 Introduction

Data extraction in the activity of team sports is a relatively new field. Through expensive systems and cameras, such as AutoStats in the NBA, professional management teams have the possibility to perform analysis of their matches, forms of play, and statistics that allow decision making for better performance. Besides, basketball coaches spend a lot of time in clipping and analyzing a match of the rival team for show to your players the tactics of the next game. To this end, it is important to provide a tool for automatically clip and extract highlights of the games. The motivation arises to reduce the prices of this kind of technology, so that small regional teams could have the possibility of analyzing their matches in a similar way -automatically- without the expense of time that represents doing it manually.

It is of interest to develop a method that, through software, allows to replace the expensive sensors that are used in the big leagues to track the players of a game, which are placed in different parts of the body, being somewhat annoying

P. Pesado and G. Gil (Eds.): CACIC 2021, CCIS 1584, pp. 95–105, 2022.
https://doi.org/10.1007/978-3-031-05903-2_7

for the athlete in his/her activity. Some examples are those that are located as girdles in certain areas such as the knee [17], chips that adhere to the athlete's body [16], or even inserted in mouthguards [12]. Microelectromechanical devices (MEMS sensors) can also be considered, which are located on the wrist of the players to allow the tracking of the players and monitor their performance [19] and [22]. Another aspect that can be addressed as a complement to the monitoring of the players, is that of the location of the ball on the field. There are different approaches today to carry out this task, such as multisensory radio frequency systems that combine player detection with ball tracking on the field [3]; radar-based systems for the detection and tracking of balls from different sports [15], or systems that use different image processing techniques to identify balls using as input the data obtained by Kinect cameras located in mobile robots [5].

The drawback found around these alternatives is that they are usually expensive and difficult to access for smaller regional teams. That is the reason why it would be useful to find an accessible way to use computer vision resources and statistical techniques to perform the game analysis of a team, such as heat maps, point counting, or the automatic extraction of clips of goal plays.

To carry out the above, it is necessary to identify the ball and track it, taking into account the moments when the ball enters the basket, which will be the goals of the match. With this action detected, it is direct to calculate the count of goals, as well as to determine start and end times in the whole video. Thus, by performing a task of recording player locations, it would be possible to generate a heat map of the team. This could allow any team to perform automatic game analysis in an accessible way, with the addition of using a non-invasive system since no sensor should be dressed during the activity.

Usually, television cameras do not perform a complete panning of the playing field, but different sectors are captured from a rotation of them. That is why to get maps of the game, it is required as a first stage the registration of the different frames of the video, and thus it is possible to generate a panoramic image that allows to visualize the complete court. Then, the result must be transformed into an aerial perpendicular view, where the visualization of the players' heat maps would be easier.

The detection of objects in images is a very incipient field, especially with the arrival of robust models of deep neural networks [2]. These models allow us to identify objects within images, which will be useful to find the players and the basketball thus obtain the data we are looking for.

This article extends the initial work presented in the CACIC conference [4] with the addition of a deep neural network for object detection, improved generation and display of heat maps over court diagrams, detection of important game events and automatic generation of short video clips for game analysis. The rest of the article is organized as follows. In Sect 2 the objectives of this work are described. Section 3 presents the details of the methods that form the proposed system. The experimental results are exposed in Sect. 4, and finally the conclusions are outlined in Sect. 5.

2   Purpose and Objectives

The main objective of this work is to develop a system that allows the extraction of clips with relevant information for the analysis of a basketball team in a match. In addition, the system should work just using a TV footage video of the game.

According to technical managers and coaches of local teams (from Santa Fe province, Argentina), it is very useful to get the distribution of the players throughout the game and, at the same time, analyze the goal situations, not only in terms of the position of the team players but also the game mode of the opponents. In order to accomplish that, the identification of the ball and the players on the field is addressed. Furthermore, if the moment when the ball enters in the basket is detected, it will be possible to extract video clips of the more relevant moments which help for a post analysis.

Once the elements of interest have been identified, the desired outputs can be generated. In particular, the distribution of players throughout a match can be visualized on a heat map, and the points can be translated into a count along with a set of small videos showing the situations, for both teams. To obtain the heat maps, it will also be necessary to study and implement image registration methods to obtain an image of the entire court, in an upper and perpendicular view.

3   The Approach

The system was designed based on a series of successive processing blocks. Figure 1 shows a general scheme of the system.

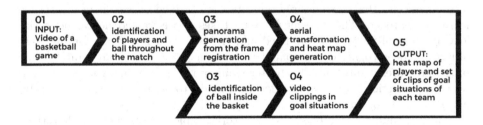

Fig. 1. Diagram of the proposed system.

The input is a TV footage video, captured from a single camera with a resolution of 1280 × 720 pixels, taken at 30 frames per second. The first step is carrying out the image registration of frames to build a panoramic image of the entire court. When registering, the transformation applied to each frame is calculated and also used to transfer the position of the players from the original frame to the panoramic view on the court at each time. For better visual interpretability, another transformation is performed that convert these results to an aerial view.

Fig. 2. Result of the identification of players (green and red boxes). (Color figure online)

Identification of Players and Ball Throughout the Match. At this stage, the YOLO object detection algorithm [2] is used. It is an algorithm based on deep neural networks trained to segment a wide variety of objects, including animals, people, among others. This network is one of the most used since through a single inference on the image, it can perform a very precise segmentation that makes it possible to implement both offline and in real time. This algorithm was trained with the Tensorflow library in Python language [1].

The publicly available YOLO network can not identify basket balls and points in the game, so these two desired outputs were trained by means of a transfer learning approach. The transfer learning takes a pre-trained neural network, leaving most of its parameters fixed and trains only a few layers for the new objects to be recognized. Compared to training a network from scratch, this process requires a substantially shorter training time, obtaining good results in terms of prediction accuracy and allowing a model to be trained with a relatively limited number of images.

The re-training was applied with a Pytorch implementation of the algorithm [11] using 942 images taken from different matches of basketball of the National Collegiate Athletic Association (NCAA). These images were labeled using the LabelImg software [18] and the results are two files: the original frame and a corresponding text file with the relative coordinates and size of the bounding boxes detected with the players and/or the point (ball inside basket situation).

The court is segmented to separate the players from the public, thus obtaining the players of any of the teams, and/or the referees. Finally, a post-processing is carried out that allows to identify which team each player belongs to, and also to discard the referees. This segmentation is carried out using different color

Fig. 3. Panoramic view generated by the image registration of the frames.

models [14]. An example of the segmentation can be seen in Fig. 2. Here, the identified players of each team are marked with boxes of different colors.

Panoramic Image Generation from Frame Registration. To generate the view of the court over which the results will be displayed, a panoramic view is first obtained from the input video [6]. To this end, an image registration algorithm was used with a step of identification of descriptors and key points detection by means of the ORB detector [9], and then a correlation search between those points using the BF Matcher [9]. After obtaining the relevant points between two successive frames, it is possible to calculate the tra3nsformation that transforms the frame from the original view to the panoramic view using a RANSAC linear estimation model [8]. All the methods were implemented in the OpenCV library [13]. An example of the final result of the registration of all the frames of the video can be seen in Fig. 3. The information calculated at this point will be relevant later to find the position of the players in this view.

Identification of Ball, Ball Inside the Basket and Video Clipping. After the transfer learning, the YOLO network is able to detect the ball through all the match, and identify when any team scores during the game, as in Fig. 4. In case of a'made basket' detection, the systems produces a short video-clip with the last 7 s and save this clip in a different registry. The algorithm detect all the frames in which the ball hit the basketball hoop, in case of missing the shot the frames detected are 2–5 approximately and in case of 'made basket' the frames count is higher than 5. In the first frame of the potential point, the algorithm takes the clip for the last 7 s and with the next frames detect if there was a score or not. An example of a moment of this situation is shown in Fig. 5.

Fig. 4. Identification of the ball.

Fig. 5. Identification of 'made basket' situation.

Aerial Transformation and Heat Map Formation. It is intended to show the heat map in an aerial view (upper and perpendicular to the court), since in this way it is possible to see the strong borders of each player, the main locations of the players, the general movement of a team, etc. To accomplish this, a transformation is made to the panoramic view of the court by placing the extreme points of the court in the panoramic view, and the destination points that respect the standard proportions of a basketball court. A perspective transformation matrix is obtained for this process [21]. The result is an aerial image of the court on which the heat maps of the match will be displayed, as can be seen in Fig. 6.

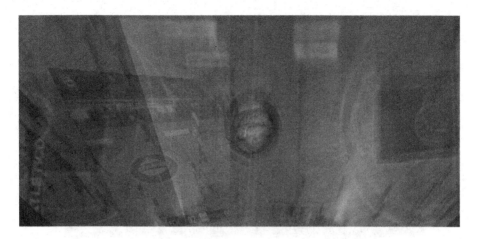

Fig. 6. Aerial view of the court.

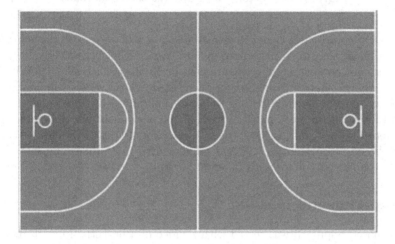

Fig. 7. Schematic representation of a basketball court.

The transformation is carried out on a panoramic image built from the average of a set of partial reconstructions obtained by the registration of different frames, so when performing the transformation to the upper view it becomes noticeable that the players were not completely eliminated from the court. While a good approximation to the top view of a basketball court is obtained, it still presents some small errors that could be improved with some non-linear noise reduction techniques (players who still appear partially).

In order to improve the detection of the edges of the court, it is an option to use its schematic representation, such as the one shown in Fig. 7. Here, a clearer visualization of the relevant information -the heat maps of the match- could be obtained, being more useful for a coach analyzing the game.

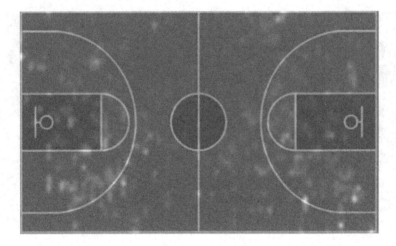

Fig. 8. Heat map for team 1 over the basketball court.

Once the coordinates that represent the location of players on the court have been obtained, the two transformations are carried out successively. The court is divided into non-overlapping square cells of 50 cm on each side [7]. With this discretization, a matrix is defined for each team, whose values will contain the number of players detected within the corresponding cells. Subsequently, with this information, the heat maps of each of the equipment are calculated using the seaborn library [20], used for the visualization of statistical data. It is open source, based on matplotlib [10] and implemented in Python, and has several models to build heat maps with different variants.

4   Experimental Results

The proposed approach was tested with a video of a basketball match recorded in 2021, between Atlético Echagüe Club and Colon S.F. Club, in the Torneo Nacional de Ascenso (Argentina).

The output images obtained after the analysis are showed in Figs. 8 and 9. These present the heat maps of each team over all the match, where each bright spot indicates presence of the team in the location and the brightest points those locations where the team has had more presence.

As can be seen, the heat map results were projected over the top view of the basketball court.

This first approach was designed and implemented to work with a typical television broad-cast video, captured with a single camera located in the upper-center of a lateral view and which only performs horizontal rotations. Respect to teams identification, it is performed using color segmentation of T-shirts and this process is parameterizable, allowing the adaptation to any teams. In addition, the basketball court detection is parameterizable, allowing its applicability to other courts.

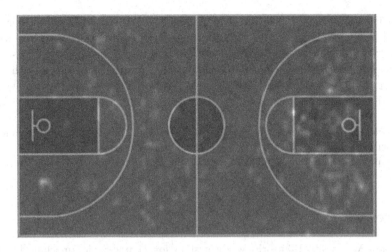

Fig. 9. Heat map for team 2 over the basketball court.

The other output provided by the system is the collection of video clips with 'goal' or 'no-goal' situations, the filenames specify each case. However, these have not information about the specific team that took the shot, then in some cases there are shots (clips) overlapped in 5 or more seconds.

Regarding the execution times, the detection of ball and basket is carried out in real time for 30FPS videos, given that the YOLO inference is very fast. The bottleneck of the performance is the creation of the heat maps, because the registration of frames only can run in real time for low resolution videos (smaller than 480p).

On the other hand, the system does not address the monitoring of individual players along the court, nor the detection failures due to occlusions. However, for the general statistics that was intended in this work, there is no significant difference.

In the registration step, a problem arises for fast game actions that took short time to go from one basket to the other. Here, the middle part of the court is poorly transformed because there is no enough frames. Nevertheless, for the general construction of the heat map, these instants could be discarded with no substantial drop in performance.

5 Conclusions

In this work, a method to obtain heat maps of the teams, and clips of goal situations, in a basketball match was presented. For the experiments, a typical television broad-cast video, captured with a single camera, was used.

Results are projected over a complete basketball court obtained through image transformations. Statistics and heat maps of players location during the match are presented as image augmentation. In addition, a set of video clips is generated resuming important moments of the match (such as the goal situation).

Future works include the improvement of the generation of the upper view of the court. In this sense, the first improvements could address to bug fixing, such as to enhance the court edges detection and to reduce noise that arises due to player shadows. Furthermore, in order to upgrade the statistics and the heat map, the ball possession and the player's position when taking a shot will be detected and analysed.

Acknowledgement. The authors would like to thank the sinc(i) institute and the UNL (with CAI+D 50620190100145LI), to Enzo Ferrante and Eric Priemer by the collaboration in the original conference work, and Ryan Werth for providing the corpus of images to train the network.

References

1. Abadi, M., et al.: {TensorFlow}: a system for {Large-Scale} machine learning. In: 12th USENIX Symposium on Operating Systems Design and Implementation (OSDI 16), pp. 265–283 (2016)
2. Bochkovskiy, A., Wang, C.Y., Liao, H.Y.M.: Yolov4: optimal speed and accuracy of object detection. arXiv preprint arXiv:2004.10934 (2020)
3. Botwicz, M., Klembowski, W., Kulpa, K., Samczyński, P., Misiurewicz, J., Gromek, D.: The concept of an RF system for the detection and tracking of the ball and players in ball sports: an extended abstract for poster presentation. In: 2017 Signal Processing Symposium (SPSympo), pp. 1–3. IEEE (2017)
4. Bourlot, J., Eberle, G., Priemer, E., Ferrante, E., Martínez, C., Albornoz, E.: Generación de mapas de calor de un partido de básquetbol a partir del procesamiento de video. In: Mac Gaul, M. (ed.) XXVII Congreso Argentino de Ciencias de la Computación, CACIC, pp. 231–239 (2021)
5. Chen, B., Huang, Z., Yu, W., Xu, Y., Peng, J.: Object recognition and localization based on kinect camera in complex environment. In: 2013 IEEE International Conference on Robotics and Biomimetics (ROBIO), pp. 668–673. IEEE (2013)
6. Chen, K., Wang, M.: Image stitching algorithm research based on opencv. In: Proceedings of the 33rd Chinese Control Conference, pp. 7292–7297. IEEE (2014)
7. Cheshire, E., Halasz, C., Perin, J.K.: Player tracking and analysis of basketball plays. In: European Conference of Computer Vision (2013)
8. Fischler, M.A., Bolles, R.C.: Random sample consensus: a paradigm for model fitting with applications to image analysis and automated cartography. Commun. ACM **24**(6), 381–395 (1981)
9. Howse, J., Minichino, J.: Learning OpenCV 4 Computer Vision with Python 3: get to Grips with Tools, Techniques, and Algorithms for Computer Vision and Machine Learning. Packt Publishing Ltd, Birmingham (2020)
10. Hunter, J.D.: Matplotlib: a 2d graphics environment. Comput. Sci. Eng. **9**(03), 90–95 (2007)
11. Jocher, G., Stoken, A., Borovec, J., Chaurasia, A., Changyu, L.: ultralytics/yolov5. Github Repository, YOLOv5 (2020)
12. Kim, J., et al.: Wearable salivary uric acid mouth guard biosensor with integrated wireless electronics. Biosens. Bioelectron. **74**, 1061–1068 (2015)
13. Laganière, R.: OpenCV 3 Computer Vision Application Programming Cookbook. Packt Publishing Ltd, Birmingham (2017)

14. Lu, W.L., Ting, J.A., Little, J.J., Murphy, K.P.: Learning to track and identify players from broadcast sports videos. IEEE Trans. Pattern Anal. Mach. Intell. **35**(7), 1704–1716 (2013)
15. Salski, B., Cuper, J., Kopyt, P., Samczynski, P.: Radar cross-section of sport balls in 0.8-40-GHZ range. IEEE Sens. J. **18**(18), 7467–7475 (2018). https://doi.org/10.1109/JSEN.2018.2862142
16. Sekine, Y., et al.: A fluorometric skin-interfaced microfluidic device and smartphone imaging module for in situ quantitative analysis of sweat chemistry. Lab on a Chip **18**(15), 2178–2186 (2018)
17. Stetter, B.J., Ringhof, S., Krafft, F.C., Sell, S., Stein, T.: Estimation of knee joint forces in sport movements using wearable sensors and machine learning. Sensors **19**(17), 3690 (2019)
18. Tzutalin, D.: Labelimg. GitHub Repository 6 (2015)
19. Wang, Y., Zhao, Y., Chan, R.H.M., Li, W.J.: Volleyball skill assessment using a single wearable micro inertial measurement unit at wrist. IEEE Access **6**, 13758–13765 (2018). https://doi.org/10.1109/ACCESS.2018.2792220
20. Waskom, M.L.: Seaborn: statistical data visualization. J. Open Source Softw. **6**(60), 3021 (2021)
21. Wen, P.C., Cheng, W.C., Wang, Y.S., Chu, H.K., Tang, N.C., Liao, H.Y.M.: Court reconstruction for camera calibration in broadcast basketball videos. IEEE Trans. Visual Comput. Graphics **22**(5), 1517–1526 (2015)
22. Zhang, X., Duan, H., Zhang, M., Zhao, Y., Sha, X., Yu, H.: Wrist mems sensor for movements recognition in ball games. In: 2019 IEEE 9th Annual International Conference on CYBER Technology in Automation, Control, and Intelligent Systems (CYBER), pp. 1663–1667 (2019). https://doi.org/10.1109/CYBER46603.2019.9066506

Software Engineering

A Sound and Correct Formalism to Specify, Verify and Synthesize Behavior in BIG DATA Systems

Fernando Asteasuain[1,2]([✉])[iD] and Luciana Rodriguez Caldeira[2]

[1] Universidad Nacional de Avellaneda, Avellaneda, Argentina
`fasteasuain@undav.edu.ar`
[2] Universidad Abierta Interamericana - Centro de Altos Estudios CAETI, CABA,
Buenos Aires, Argentina
`luciana.rodriguezcaldeira@alumnos.uai.edu.ar`

Abstract. In this work we consolidate our behavioral specification framework based on the Feather Weight Visual Scenarios (FVS) language as a powerful tool to specify, verify and synthesize behavior for BIG DATA systems. We formally demonstrate that our approach is sound and correct end to end, including the latest extensions such as fluents and partial specifications. In addition, our empirical validation is strengthen by adding new and complex case studies and incorporating, besides execution time, space exploration as a factor in the comparison with other approaches. We believe that the contributions introduced in this work aim to point up FVS as a solid tool to formally verify behavior in BIG DATA syst

Keywords: Formal verification · BIG DATA · Soundness · Correctness

1 Introduction

Modern world is surrounded by interconnected software devices and artifacts. Almost every device is equipped with some kind of sensor capable of transmitting or receiving information. The topologies and ways to communicate each other are as variable as the kind of domains they can be applied. New trends and technologies arise such as Industry 4.0 [38] or the so called Internet of Things [31]. However, perhaps the most relevant cutting edge emergent topic is built around the concept of BIG DATA systems [24]. These systems include the process of gathering and analyzing the possible huge amount of information available. It is said that the main features concerning BIG DATA systems can be summed up with the "Five Vs": Variety, Velocity, Volume, Value and Veracity [19,33].

The software engineering community had been trying to address the challenges involved in BIG DATA systems [7,12,18,26–29,35,36] . That is, providing rigourous and sound methodologies, tools, processes and techniques to carry on the complex process of manipulating information in BIG DATA Systems. One of the fields that need particular attention is formal verification [21,28]. According

© The Author(s), under exclusive license to Springer Nature Switzerland AG 2022
P. Pesado and G. Gil (Eds.): CACIC 2021, CCIS 1584, pp. 109–123, 2022.
https://doi.org/10.1007/978-3-031-05903-2_8

to [28] only 2 of nearly 200 analyzed approaches handling software engineering and big data are related to this particular area. Most of these approaches are focused on just two of the five V's: Velocity and Volume, trying to improve the performance and size exploration of the system under analyses making available, for example, parallel or distributed versions of their tools.

However, the other three V's (Variety, Value and Veracity) have been somehow neglected. These three aspects need powerful, rich and expressive formalisms and notations to reason about the expected behavior of this kind of systems. This is because they need to handle heterogeneous and unstructured data. In addition, proper software engineering approaches need to be proved sound and correct [22,23,34].

In [3] we presented the Feather Weight Visual (FVS) framework as a candidate tool to formally apply verification in BIG DATA systems contemplating all the five V's. For Velocity and Volume, we parallelised our scheme and combined FVS with the parallel version of the MTSA model checker(Modern Transition System analyzer), which can be downloaded from https://mtsa.dc. uba.ar/. Regarding Variety, Value and Veracity we extended FVS with powerful formalisms: fluents [25], partial specifications with Modal Transition Systems [30] and controller synthesis [13]. Fluents are a compelling logic which introduce a new layer of abstraction in the specification, creating and reasoning with high level events occurring between intervals of time. Partial specifications introduces the novelty of optional behavior which can be latter decided to be included or not in future versions, constituting an appealing tool to address the description of behavior in early stages. Finally, controller synthesis is a potent technique to build a system such that is satisfies its specification by construction. This is achieved employing game theory algorithms.

We now present a robust continuation of that work from two crucial perspectives: theoretical and applied. From the theoretical point of view we provide formal proofs of our approach. In particular, we formally demonstrate that FVS are capable of dealing fluents and partial specifications, the two missing links needed to state that FVS is sound and correct all the way, from the beginning to the end. From the applied perspective, we exhibit a solid empirical validation considering two case of studies available in the literature, and a comparison with other attractive approach [12]. The empirical evaluation considers not only execution time, but also space size exploration. Size and time are by far the most relevant issues in formal verification algorithms. In addition, all the source files from the case studies are available online so that results can be reproduced and replicated.

1.1   Previous Work and New Contributions

FVS was firstly introduced to specify behavior in early stages [1]. Among the distinguishable features of this language we can mention a powerful and expressive notation, being more expressive than classical temporal logics or state-based formalisms [1]. FVS was latter extended to denote branching behavior besides its original linear flavour, coping with the most classic way to reason about the expected behavior of systems [4]. FVS continue to growth incorporating the possibility to synthesize behavior [5]. That is, FVS can not only describe behavior but also a concrete implementation can be automatically build obtaining a controller for the system under analysis.

FVS was then adapted to address with formal verification of BIG DATA systems. The first step was given in [2], where we present a parallel algorithm that translates FVS specifications into Büchi automata. The formal demonstration of the correctness and soundness of the parallel approach is given in [6]. Finally, in [3] we deepened this road by composing FVS parallel translation algorithm with a parallel version of the MTSA model checker tool [15]. In order to achieve this combination FVS was extended using fluents [25] and partial specifications [30]. We now complete this last step presenting novel and valuable contributions:

1. We formally demonstrate that FVS can denote all the behavior expressible by fluents.
2. We formally demonstrate that FVS can handle partial specifications. To do so, we proved that FVS semantics can be seen as a refinement operation between rules.
3. The formal proofs of our approach constitute a key aspect regarding the correctness and soundness of our approach.
4. We provide a more profound and enriched empiric validation of our scheme. Complete, relevant case of studies are introduced, comparing our approach with a widely known tool in the literature [12]. Besides considering execution time, we incorporated size as a comparison factor. In addition, all the examples' source files and our tool are provided online, making reproducible the results (see https://gitlab.com/fernando.asteasuain/fvsweb).

The rest of this paper is structured as follows. Section 2 introduces the FVS framework. Sections 3 and 4 exhibit how FVS can handle fluents and partial specifications respectively, including the formal demonstration to prove our approach to be sound and correct. Sections 5 and 6 cover the case studies whereas Sect. 7 condenses all the empiric validation results and analysis. Section 8 addresses related and future work and Sect. 9 presents the conclusions of this work.

2   Feather Weight Visual Scenarios

In this section we will informally describe the main characteristics of FVS whereas the complete definition of the language can be found in [1].

FVS is a graphical language based on scenarios. Scenarios describe partial order of events through the use of labeled points and labeled arrows. Points are labeled with the events that occur on that particular moment. In particular, points can be labeled with a logic formulae. Arrows between points indicates a precedence relationship. For instance, in Fig. 1-(a) A-event precedes B-event. In FVS it can be specified the very next occurrence of an event and not any other occurrence in the future. This is graphically shown with a small arrow near the point of interest. For example, Fig. 1-b shows an scenario that properly signal the first occurrence of a B event after the occurrence of a A event. Arrows can be labeled to restrict behavior. Figure 1-c exhibits a situation where a C event must not occur between events A and B. In FVS, aliasing between points can also be denoted. Scenario in 1-d indicates that a point labeled with A is also labeled with $A \wedge B$. Note that A-event is repeated on the labeling of the second point just because of FVS formal syntaxis. This constitute an interesting syntactic sugar mechanism of our notation.

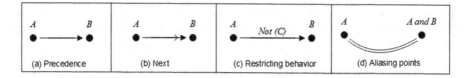

Fig. 1. Basic elements in FVS

We now introduce the concept of FVS rules, a core concept in the language. FVS rules consists of one scenario playing the role of an antecedent or a triggering condition, and one or more consequents, playing the role of the behavior to be triggered after the condition is met. The intuition is that whenever a trace "matches" a given antecedent scenario, then it must also match at least one of the consequents. Graphically, the antecedent is shown in black, and consequents in grey. Since a rule can feature more than one consequent, elements which do not belong to the antecedent scenario are numbered to identify the consequent they belong to. Some examples are shown in Fig. 2 describing the behavior of a semaphore controlling access to a critical section in a distributed environment. The rule at the top says that once a process is in the critical section no other process will enter until it exits. The second rule (the one in the middle) has two consequents. After an *Critical* event a process will either enter the critical section (Consequent 1) or it will be moved to the waiting queue (Consequent 2). Finally, rule at the bottom of Fig. 2 introduces a fairness concern. It says that between two consecutive admission of a process *P1* to the critical section, if another process *P2* requires to enter, its access will be satisfied. This avoids that process *P2* dies from starvation, a typical concern in distributed systems.

2.1   FVS and Ghosts Events

Besides the actual events occurring in the system, FVS is able to reason about high-level events [1]. Adding new layers of abstraction is a common and useful practice when specifying behavior. We denote this high-level events as "ghost" events while system's events are called "actual" events. We employ FVS rules to model how "ghost" events are activated, and latter, other FVS rules expressing behavior with these events. In this way, FVS can introduce a new layer of abstraction. For example, in a software architecture system one way to address

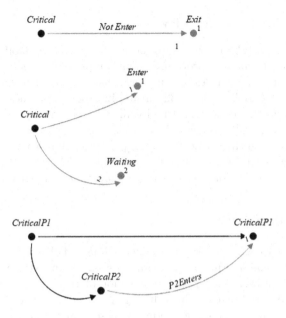

Fig. 2. FVS rules examples

the availability concern is detecting when a server is near to its maximum capacity and reduces the level of services to offer to avoid a server's crash down. This downgrade mode in the server can be seen as a ghost event and occurs when the systems detects that the number of connection is too high. The first rule in Fig. 3 establishes the conditions that triggers the occurrence of the ghost event *DowngradeMode*, whereas the second rule directly shapes behavior using ghosts events. It says that while working in the down grade mode the server reduces the services it can respond until the downgrade is over.

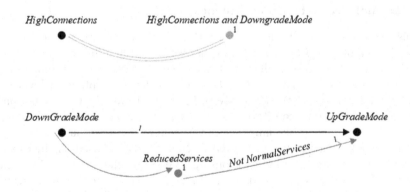

Fig. 3. An FVS ghost event behavior

2.2   Behavioral Synthesis in FVS

FVS specifications can be used to automatically obtain a controller employing a classical behavioral synthesis procedure. The complete process is fully described in [5]. In few words, FVS rules are first translated into Büchi automata. Then we relied on several tools such as [32] or [37] in order to synthesize the denoted behavior an obtaining a controller. Finally, in [3] we showed how FVS can also be combined with the parallel version of the MTSA model checker [15]. Obtaining a controller complete all the verification path point to point, from the specification to the implementation. In this way, we can state that FVS is a proper framework to handle system's behavior specification, verification and synthesis.

3   Fluents and FVS

In this section we explain how FVS can denote fluents behavior, and we formally demonstrate this claim. Fluents [25] can be seen as an extension of the LTL logic. A fluent represent a particular behavior of the system that take place between intervals or moments, introducing a new layer of abstraction in the specification. Starting and ending actions of these intervals are defined, and then properties can be stated using the mentioned intervals. For example, recall the FVS ghost event example in Fig. 3. This downgrade mode can be specified using a fluent defining the *HighConnection* event as an starting event, a *LowConnection* event as a ending event, and a LTL formula such as □ (DowngradeMode ⇒ *ReducedServices*).

FVS can specify fluents in a very simple and direct way employing ghosts events. Fluents starting and ending delimiters are modeled with FVS rules, fluents predicates are modeled with ghost events, and intervals behavior are simply FVS rules using those ghosts events. The next subsection addresses the formal proof showing that FVS can denote all behavior expressible in fluents.

3.1   Formal Proof: FVS and Fluents

In this section we formally demonstrate that FVS can denote all the behavior expressible employing Fluents [25]. In other words, given a fluent FL we build a set $RFVS$ of FVS rules such that the language denoted by FL is included in the language denoted by $RFVS$: $L(FL) \subseteq L(RFVS)$. A few concepts must be cleared before the formal demonstration is given. First of all, we adopted the notion of traces as a linear sequence of events, and those traces accepted by the fluents or the FVS rules defines the language of the system. More formally, a trace t is seen as the sequence $\{t_0, t_1, \ldots, t_k\}$. FVS semantics [1] is based on the concept of morphisms, a mathematical relationship that establishes when a given trace satisfy an FVS rule. In few words, if for every item in a trace t a morphism m exists such that t can be related to a rule R, then t satisfies R. More formally, $\forall i, 0 \leq i \leq k$ in $\{t_0, t_1, \ldots, t_k\}$, there exists a morphism m_i such that $t_i \xrightarrow{m_i} t_{i+1}$. It can be said that each morphism makes an advance in the trace.

A fluent is formally defined establish a set of initiating actions, a set of terminating actions, and the behavior itself happening between the start and the end [25]. That is, a fluent $FL = < I_{FL}, T_{FL} >$, where I_{FL} denotes the starting events, T_{FL} the ending events, and $I_{FL} \cap T_{FL} = \emptyset$. In addition, a boolean variable called *Initially* is defined to express if the fluent is true or false at the beginning of the execution. Finally, a trace $t = \{t_0, t_1, \ldots, t_k\}$ is said to be valid for a fluent if one of the following conditions holds:

1. Condition 1: *Initially* $\wedge$ ($\forall i, 0 \leq i \leq k, t_i \not\exists T_{FL}$). This conditions covers the case where the fluent is valid in the beginning and no terminating events occurred.
2. Condition 2: $\exists j, j \leq k, t_j \in I_{FL} \wedge (\forall i, j \leq i \leq k, t_i \not\exists T_{FL})$. This condition covers the case where the fluent initiates at a given point, and no terminating events occurs.

We are now ready to address the formal demonstration showing that FVS can denote any behavior expressible by a fluent. Given a fluent FL we define a set of FVS rules called $RFVS$ in the following way:

1. For every action a in I_{FL} there exists a rule r_a in $RFVS$. That is, for every initiating event there is an FVS rule that starts the behavior of the fluent.
2. For every action a in T_{FL} there exists a rule r_a in $RFVS$. That is, for every terminating event there is an FVS rule that ends the behavior of the fluent.
3. For every action established for the fluent, we add a rule in $RFVS$ mimicking its behavior.
4. If the variable *Initially* is set to true, we add an FVS rule activating the fluent.

We now demonstrate that the language of a fluent FL is included in the language denoted by $RFVS$: $L(FL) \subseteq L(RFVS)$. That is, given a trace $t = \{t_0, t_1, \ldots, t_k\}$ active in the fluent FL we will show that $t \in L(RFVS)$.

If t is active in FL then either Condition 1 or Condition 2 must hold. Let's assume that Condition 1 holds. Since FL's *Initially* variable is true, there exists a rule r in $RFVS$ initiating the fluent behavior. So, there exists a morphism m_0 such that $t_0 \xrightarrow{m_0} t_1$. What is more, since $\forall i, 0 \leq i \leq k, t_i \not\exists T_{FL}$, the fluent will always advance (no ending events occurs). The set of rules $RFVS$ includes, by construction, all the rules that advances a fluent. This implies that $\forall i, 1 \leq i \leq k$, there exists a morphism m_i such that $t_i \xrightarrow{m_i} t_{i+1}$. Therefore, the trace t belongs to the set of traces denoted by $RFVS$: $t \in L(RFVS)$, which was the goal of our demonstration. $\square$.

Now let's analyze the case where Condition 2 holds. If Condition 2 holds, then there exists an event j which activates the fluent. Given the way the set of rules $RFVS$ was build (including one rule for each activation), we can affirm that there exists a morphism m_j such that $ti \xrightarrow{m_j} t_j$. Since no terminating events occurs, there will always be a next morphism advancing the trace: $\exists$ morphism m_n such that $t_n \xrightarrow{m_n} t_{n+1}$, $n \leq j \leq k$. Therefore, $t \in L(RFVS)$. This concludes the second part of the demonstration. $\square$.

Since we have proved that the hypothesis holds for both conditions, we can conclude that $L(FL) \subseteq L(RFVS)$.□ □.

4 Partial Specifications and FVS

Partial Specifications are a crucial tool to model and shape early behavior of computer systems, introducing the possibility to distinguish between optional and required behavior. Probably the most widely known formalism addressing Partial Specifications is Modal Transitions Systems (MTS) [30]. A process called *refinement* guides the iterations between specifications versions, guaranteing that the new version satisfy all the original assumption in the system.

FVS includes the possibility to express partial and required behavior since FVS rules can hold multiple consequents. So, a future rule modifying the number of consequents is a refinement of the previous version. So, FVS provides the refinement operation by its traces semantics definition. A rule with multiple consequents can be replaced in next versions with a rule with less consequents (the optional behavior is discarded), or combining two consequents into one (making mandatory an optional behavior). In the following subsection we formally demonstrate this claim.

4.1 Formal Proof: FVS and Partial Specifications

In this section we formally demonstrate that FVS rule semantics can be seen as a refinement relationship, making FVS a suitable notation to denote partial specifications. For the formal demonstration we employed MTS [30], since this formalism is one of the most widely known addressing partial specifications. The most distinguishable feature in partial specifications is the possibility to divide events into two categories: required and optional behavior.

The refinement operation relates two different versions of the system, where the newer version makes former optional behavior either turned into required or discarded. More formally, given two MTS M and N we define a refinement relationship H, $H \subseteq States(M)$ X $States(N)$ if and only if, $\forall\ l \in events(M) \land \forall$ (s,t) $\in H$, the following conditions holds:

1. if $(s,l,s') \in Required(M)$ then $\exists\ t$ such that $(t,l,t') \in Required(N) \land (s',t') \in H$.
2. if $(t,l,t') \in Optional(N)$ then $\exists\ s$ such that $(s,l,s') \in Optional(M) \land (s',t') \in H$.

In other words, we can establish that an MTS N refines another MTS M if every required transition in M belongs to N and every optional behavior of N is present in M.

The hypothesis to be demonstrated is that an FVS rule with multiple consequents represent a valid refinement operation. That is, given an FVS rule M with k consequents, and an FVS rule N with k-1 consequents, where $Consequents(M) = Consequents(N) \cup C_k$, $\land\ C_k$ is the missing consequent in N. Given this setting, we must prove that the traces accepted by N are a refinement operation for the traces accepted by M. According to the refinement operation definition, we must prove the following claims:

1. $\forall$ trace $t \in$ *required(M)*, $t \in$ traces(N)
2. $\forall$ trace $t \in$ *optional(N)*, $t \in$ traces(M)

To prove Claim 1 we consider two cases: where t satisfies the rule and where t does not satisfy the rule. If the trace t does not satisfy rule M, t does not satisfy rule N either, since N is the same rule without one consequent. Since traces that do not satisfy the rules are not included in the system by definition of FVS semantics, this case can be discarded. Therefore, we must conclude that t do satisfy M. This indicates that given $t = \{t_0, t_1, \ldots, t_k\}$, there exists a morphism $m_i \; \forall \; i$, $0 \leq i \leq k$ such that $t_i \xrightarrow{m_i} t_{i+1}$, leading t to an accepting state. Let C_k be the missing consequent in N (recall that *Consequents(M) = Consequents(N)* $\cup \; C_k$). Lets suppose by the absurdum that $t \notin$ traces(N). This implies that one of the morphisms m_i belongs exclusively to consequent C_k. However, this indicates that t does not satisfy the rule, which contradicts the initial assumptions. The contradiction arose from assuming that $t \notin$ *traces(N)*, therefore $t \in$ *traces(N)*, which was what we aimed to prove. This concludes the first part of the demonstration. $\square$.

We now address Claim 2: $\forall$ trace $t \in$ *optional(N)*, $t \in$ traces(M). As we did with Claim 1, the case where t does not satisfy the rule can be ignored. Therefore, we can assume that t satisfy the rule. Let assume by the absurdum that $t \notin$ traces(M). If that is the case, then there exists a morphism m_i such that $t_i \xrightarrow{m_i} t_{i+1}$ for some $i \wedge m_i$ belongs exclusively to C_k. However, if this is true, then t does not satisfy the rule, which is a contradiction of the initial assumptions. And this contradictions backtracks to assuming that $t \notin$ traces(M). This imply that $t \in$ traces(M), which was our goal to be proved. This concludes the second part of the demonstration. $\square$.

Given that Claim 1 and Claim 2 had been proved, we can affirm that FVS rules denote a refinement relationship. $\square$ $\square$.

5   Simple Load Balancing Case Study

Inspired on [12] we completely formalized the Simple Load Balancing example, including the specification of its behavior and the automatic construction of a controller for it. The setting environment for the load balancer system consists of c clients, two servers, and a load balancer process playing the role of a proxy between them. As in [12], we studied a model with 2, 5 and 10 clients. Some rules describing the behavior can be found in Fig. 4. The rule in Fig. 4-a says that every client should eventually make a request whereas the rule in Fig. 4-b states that the server enters the idle state when it finishes processing a client's request. Finally, rule in Fig. 4-c dictates that the server can not enter the idle state while processing a request.

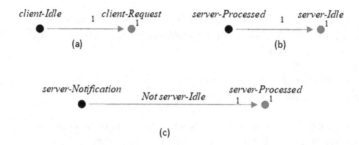

Fig. 4. Some FVS graphical rules for the load balancer example

Part of the controller for the Load Balancer Server Controller is shown in Fig. 5. Recall the complete files can be inspected from the FVS web site (https://gitlab.com/fernando.asteasuain/fvsweb), including both the server and client controllers.

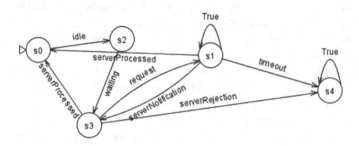

Fig. 5. Part of the load balancing controller system for the server side

6   Shared Memory Case Study

Completing the empiric evaluation presented in [12] we now present the FVS specification for a shared memory example. This system represents clients trying to gain access to a shared memory through a unique bus. As in [12], we specified two variations of the system: with 5 clients and with 10 clients.

Three rules describing the behavior can be found in Fig. 6. The first one (see Fig. 6-a) establishes a progress property: every process must advance to the active state while waiting in the queue. The property in Fig. 6-b demands that processes must be in memory in order to be in the active state. Finally, rule in Fig. 6-c says that if a process should wait in the queue if the intermediate memory is full.

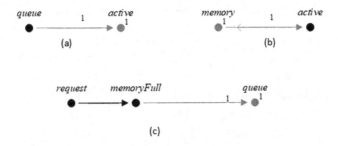

Fig. 6. Three FVS rules for the shared memory example

Part of the controller for the Shared Memory example is shown in Fig. 7.

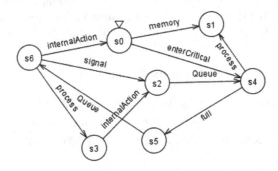

Fig. 7. A controller for the shared memory system

7 Empirical Evaluation

In this section we present a comparison between our approach and the technique presented in [12], which also addresses formal verification in big data systems and we conducted the same case of studies. We compared not only execution times but also the space dimension. So, both views, space and time, which are critical in any formal approach are thoroughly analyzed. Although they verified the behavior of the algorithms and we obtained controllers instead, the involved tasks and objectives are similar enough to produce valuable results from their performance execution time and size comparison. We ran our experiments in a Bangho Inspiron5458, with a Dual Core i5-5200U and 8 GB RAM memory.

Table 1 shows the empirical comparison considering the execution time view. The column MAP-Reduce-CTL stands for the technique in [12] while the column ParallelFVS stands for our approach. Execution times are expressed in seconds.

Table 1. Empirical comparison: execution time

Example	Map-Reduce CTL	Parallel FVS
5-Shared	110 s	188 s
10-Shared	1032 s	1341 s
2-Load balancing	43 s	76 s
5-Load balancing	114 s	136 s
10-Load balancing	18958 s	20022 s

Table 2 shows the empirical comparison considering the size dimension. In this case, we present the results for two concrete examples: the Shared Memory example with 5 clients and the Load Balancing example with 5 clients. Similar results were obtained for the other versions of the examples considering more clients. For measuring size, we considered the number of states (st in Table 2) and transitions (tr in Table 2) of the automata involved in the verification process.

Table 2. Empirical comparison: size

Example	Map-Reduce CTL	Parallel FVS
5-Shared	1863 st 7528 tr	3600 st 6500 tr
5-Load balancing	116175 st 425604 st	301001 st 432653 tr

Valuable observations can be stated by analyzing both tables. Considering time execution, it can be observed that our approach holds a worst execution time compared with [12] in small examples. We believe the reason behind this is that the parallel overhead in our approach overcomes the parallel benefits. However, the difference diminishes as the examples become more complex, being almost negligible for the most complex cases (for example, the shared memory with 10 clients or the load balancing example with 15 clients). This is a good result for our approach, since we can offer our rich and expressive specification language without resigning performance.

Regarding the size comparison (see Table 2) it can be seen that the automata obtained in our approach are bigger, specially considering the number of states. FVS approach almost duplicate the number of states while the number of transitions are pretty similar. Our main hypothesis to explain these numbers is that the algorithm translating FVS rules into Büchi automata is introducing more states than the ones that are actually needed. This particular point should be addressed in future work. However, as seen in Table 1 this size difference does not has an actual impact in the execution times.

In conclusion, we can state that FVS came out unhurt from the empirical validation point of view. FVS can offer a powerful and rich notation to express and synthesize behavior in a competitive manner considering execution time and size of the involved artifacts.

8   Related and Future Work

Perhaps the pioneer approach addressing formal verification in BIG DATA systems is [12,18]. This work is focused on distributed CTL (computation tree logic) and the architecture of their scheme is based on a HADOOP MAPREDUCE fashion, a classic architecture in this domain. The distributed state space exploration rely on a framework called Mardigras [11]. This latter tool is able to work with very different formalisms such as temporal logics or Petri Nets. Given this tool flexibility we would definitively like to explore if FVS can be integrated with Mardigras since the potential benefits of this combination sounds promising.

Other valuable approaches such as [9,10,14,16,17] focused on parallel or distributed model checking. It would be interesting to check if the integration of one of this tools with FVS outcomes the results obtained with FVS and MTSA presented in this paper. Similarly, other approaches aim to speed-up the model checking by performing parallel verification of very small units pieces of behavior [8,20] . For example, these units are called *swarms* in [20].

Regarding future work, we would like to strengthen our empirical validation by exploring the interaction of FVS with other model checkers. Another future research line is trying to optimize the algorithm translating FVS scenarios into Büchi automata. As noted in Sect. 7 the automata built by our algorithm do present more states than other known tools. Despite this does not have an impact on performance is definitively an issue to be addressed. Finally, we would like to add expressiveness as a factor to compare FVS with other approaches.

9   Conclusions

In this work we accentuated FVS as an appealing framework to address formal verification aspects in BIG DATA domains. From the theoretical point of view, we incorporated formal proofs for our approach, guaranteeing the soundness and correctness of all the involved steps from end to end. From the applied point of view, we analyzed a complete set of benchmarks and compare FVS with the most successful tool in the field. We believe that the theoretical and applied contributions introduced in this work are relevant for the formal verification software engineering phase in a forefront domain such as BIG DATA systems.

References

1. Asteasuain, F., Braberman, V.: Declaratively building behavior by means of scenario clauses. Requirements Eng. **22**(2), 239–274 (2016). https://doi.org/10.1007/s00766-015-0242-2
2. Asteasuain, F., Caldeira, L.R.: A parallel tableau algorithm for big data verification. In: CACIC, pp. 360–369 (2020). ISBN 978-987-4417-90-9

3. Asteasuain, F., Caldeira, L.R.: An expressive and enriched specification language to synthezise behavior in big data systems. In: CACIC, pp. 357–366 (2021). ISBN 978-987-633-574-4

4. Asteasuain, F., Calonge, F., Gamboa, P.: Behavioral synthesis with branching graphical scenarios. In: CONAIISI (2019)

5. Asteasuain, F., Federico, C., Manuel, D., Pablo, G.: Open and branching behavioral synthesis with scenario clauses. CLEI J. **24**(3), 1–20 (2021)

6. Asteasuain, F., Luciana, R.C.: Exploring parallel formal verification of big-data systems. Revista Ciencia y Tecnología **21**(2), 7–18 (2021)

7. Balouek, D., et al.: Adding virtualization capabilities to the grid'5000 testbed. In: Ivanov, I.I., van Sinderen, M., Leymann, F., Shan, T. (eds.) CLOSER 2012. CCIS, vol. 367, pp. 3–20. Springer, Cham (2013). https://doi.org/10.1007/978-3-319-04519-1_1

8. Barnat, J., Bauch, P., Brim, L., Češka, M.: Employing multiple CUDA devices to accelerate LTL model checking. In: 2010 IEEE 16th International Conference on Parallel and Distributed Systems, pp. 259–266. IEEE (2010)

9. Barnat, J., Brim, L., Češka, M., Ročkai, P.: Divine: parallel distributed model checker. In: 2010 ninth PDMC, pp. 4–7. IEEE (2010)

10. Bell, A., Haverkort, B.R.: Sequential and distributed model checking of petri nets. STTT J. **7**(1), 43–60 (2005)

11. Bellettini, C., Camilli, M., Capra, L., Monga, M.: Mardigras: simplified building of reachability graphs on large clusters. In: RP Workshop, pp. 83–95 (2013)

12. Bellettini, C., Camilli, M., Capra, L., Monga, M.: Distributed CTL model checking using map reduce: theory and practice. CCPE **28**(11), 3025–3041 (2016)

13. Bloem, R., Jobstmann, B., Piterman, N., Pnueli, A., Sa'Ar, Y.: Synthesis of reactive (1) designs (2011)

14. Boukala, M.C., Petrucci, L.: Distributed model-checking and counterexample search for CTL logic. IJSR **3,3**(1–2), 44–59 (2012)

15. Brassesco, M.V.: Síntesis concurrente de controladores para juegos definidos con objetivos de generalized reactivity(1). Tesis de Licenciatura., http://dc.sigedep.exactas.uba.ar/media/academic/grade/thesis/tesis_18.pdf UBA FCEyN Dpto Computacion (2017)

16. Brim, L., Černá, I., Moravec, P., Šimša, J.: Accepting predecessors are better than back edges in distributed LTL model-checking. In: FMCAD, pp. 352–366 (2004)

17. Brim, L., Yorav, K., Žídková, J.: Assumption-based distribution of CTL model checking. STTT **7**(1), 61–73 (2005)

18. Camilli, M.: Formal verification problems in a big data world: towards a mighty synergy. In: ICSE, pp. 638–641 (2014)

19. Cappa, F., Oriani, R., Peruffo, E., McCarthy, I.: Big data for creating and capturing value in the digitalized environment: unpacking the effects of volume, variety, and veracity on firm performance. J. Prod. Innov. Manage. **38**(1), 49–67 (2021)

20. DeFrancisco, R., Cho, S., Ferdman, M., Smolka, S.A.: Swarm model checking on the GPU. STTT **22**(5), 583–599 (2020)

21. Ding, J., Zhang, D., Hu, X.-H.: A framework for ensuring the quality of a big data service. In: 2016 SCC, pp. 82–89. IEEE (2016)

22. Drechsler, R., et al.: Advanced Formal Verification. vol. 122. Springer, New York (2004). https://doi.org/10.1007/b105236

23. D'silva, V., Kroening, D., Weissenbacher, G.: A survey of automated techniques for formal software verification. IEEE Trans. Comput. Aided Des. Integr. Circuits Syst. **27**(7), 1165–1178 (2008)

24. Emani, C.K., Cullot, N., Nicolle, C.: Understandable big data: a survey. Comput. Sci. Rev. **17**, 70–81 (2015)
25. Giannakopoulou, D., Magee, J.: Fluent model checking for event-based systems. In: European Software Engineering Conference, pp. 257–266 (2003)
26. Hummel, O., Eichelberger, H., Giloj, A., Werle, D., Schmid, K.: A collection of software engineering challenges for big data system development. In: SEAA, pp. 362–369. IEEE (2018)
27. Inverso, O., Trubiani, C.: Parallel and distributed bounded model checking of multi-threaded programs. In: PPoPP, pp. 202–216 (2020)
28. Kumar, V.D., Alencar, P.: Software engineering for big data projects: domains, methodologies and gaps. In: 2016 IEEE International Conference on Big Data (Big Data), pp. 2886–2895. IEEE (2016)
29. Laigner, R., Kalinowski, M., Lifschitz, S., Monteiro, R.S., de Oliveira, D.: A systematic mapping of software engineering approaches to develop big data systems. In: SEAA, pp. 446–453. IEEE (2018)
30. Larsen, K.G., Thomsen, B.: A modal process logic. In: LICS, pp. 203–210. IEEE (1988)
31. Li, S., Xu, L.D., Zhao, S.: The internet of things: a survey. Inf. Syst. Front. **17**(2), 243–259 (2014). https://doi.org/10.1007/s10796-014-9492-7
32. Maoz, S., Ringert, J.O.: Synthesizing a Lego forklift controller in gr (1): a case study. arXiv preprint arXiv:1602.01172 (2016)
33. Naeem, M., et al.: Trends and future perspective challenges in big data. In: Pan, J.-S., Balas, V.E., Chen, C.-M. (eds.) Advances in Intelligent Data Analysis and Applications. SIST, vol. 253, pp. 309–325. Springer, Singapore (2022). https://doi.org/10.1007/978-981-16-5036-9_30
34. Nejati, S.: Next-generation software verification: an AI perspective. IEEE Softw. **38**(3), 126–130 (2021)
35. Otero, C.E., Peter, A.: Research directions for engineering big data analytics software. IEEE Intell. Syst. **30**(1), 13–19 (2014)
36. Sri, P.A., Anusha, M.: Big data-survey. Indonesian J. Electr. Eng. Inf. (IJEEI) **4**(1), 74–80 (2016)
37. Tsay, Y.-K., Chen, Y.-F., Tsai, M.-H., Wu, K.-N., Chan, W.-C.: GOAL: a graphical tool for manipulating Büchi automata and temporal formulae. In: Grumberg, O., Huth, M. (eds.) TACAS 2007. LNCS, vol. 4424, pp. 466–471. Springer, Heidelberg (2007). https://doi.org/10.1007/978-3-540-71209-1_35
38. Xu, L.D., Xu, E.L., Li, L.: Industry 4.0: state of the art and future trends. Int. J. Prod. Res. **56**(8), 2941–2962 (2018)

Data Variety Modeling: A Case of Contextual Diversity Identification from a Bottom-up Perspective

Líam Osycka[(✉)], Agustina Buccella[ID], and Alejandra Cechich[ID]

GIISCO Research Group - Departamento de Ingeniería de Sistemas,
Facultad de Informática, Universidad Nacional Del Comahue, Neuquen, Argentina
liam.osycka@fi.uncoma.edu.ar

Abstract. *Variety* is a property related to data diversity in Big Data Systems (BDS) that comprises several cases, such as structural diversity (variety in data types), source diversity (variety in the way data are produced), etc. Recently, adding contextual information allows more complex analyses, which open the possibility of modeling variety thinking of reuse. In this article, we introduce a proposal for modeling variety by following a dual data processing perspective. We exemplify the case of one of these perspectives by identifying and modeling contextual variations in a particular domain problem.

Keywords: Big data systems · Variety modeling · Software product lines

1 Introduction

Currently, adding semantics to Big Data systems (BDS) is an increasing tendency for architectural properties, such as interoperability, security, reusability, etc. [9].

Reusability has been approached from diverse perspectives in BDS development. For example, in [11] reuse concepts are discussed in the context of data analytics distinguishing among data use/reuse. Open research questions about reusability typically focus on trade-offs analyses between collecting new data and reusing existing ones, and managerial aspects for data reuse. More specifically, in [5] data privacy aspects are analyzed in a similar sense. Here, a taxonomy for data reuse is proposed to determine whether this reuse should be allowed and the conditions that would ensure privacy preservation.

Other proposals focus on aspects of reuse in terms of increasing collaboration when developing BDS by using new technologies: for instance, in [12] a development approach using storage and processing capabilities of a public cloud is introduced. Additionally, different platforms for developing BDS are approached from the perspective of reuse; for instance in [1] improvements in efficiency of

P. Pesado and G. Gil (Eds.): CACIC 2021, CCIS 1584, pp. 124–138, 2022.
https://doi.org/10.1007/978-3-031-05903-2_9

tools, such as Apache Hadoop[1] and Spark[2], are analyzed by identifying common aspects when reusing artifacts among different projects.

Similarly, by including the identification of common and variant aspects as system families, in [7] a reference architecture is mapped to use cases (strategic geospatial information analysis and visualization, signal analysis, etc.). From these cases, relevant requirements are identified, including categories such as data types (unstructured text, geospatial, audio, etc.), data transformation (clustering, correlation), data visualization (images, networks), etc. The architecture is organized as a collection of modules that decompose the solution into elements representing functions or capabilities of a set of aspects (external requirements, reusable modules, roles, stakeholders, commercial and open source of-the-shelf data, etc.). Finally, the modules are grouped into three categories: (1) Big data application provider (including business cases for the system), (2) Big data framework provider (including platforms, storage, etc.), and (3) Transversal modules (including different aspects for developing BDS).

Differently from this last proposal, ours is built upon variability modeling in software product lines, explicitly adding variety identification. Trying to answer the research question *RQ: How can we identify data variety and model reusable elements in Big Data system developments?*, our work starts from a data processing pipeline that incorporates data variety by modeling common and variant features as variability of a software product line (SPL). Depending on the perspective of analysis, data may be approached in a top-down or bottom-up processing perspective.

In a previous work [10], we have introduced the bottom-up perspective at a conceptual level, and exemplified its use through a motivating example. In this paper, we extend this work by: (1) introducing both, the top-down and the bottom-up perspectives, as step-wise procedures; and (2) elaborating the example, including implementation decisions, a more extensive description, and a deeper discussion of its results. In addition, we include a previous work section for an easy understanding of our particular software product line modeling.

This paper is organized as follows. The next section introduces some techniques we have previously designed to deal with variability modeling. Then, Sect. 3 introduces both perspectives of our proposal, and Sect. 4 discusses a case of contextual diversity from the bottom-up perspective.

2   Previous Work

Our approach emerges as a functionality-oriented design [2,3]. That is, each functionality of the SPL is documented by a functional datasheet representing the set of services, commons[3] and variants, which interact to reach the desired functionality. Each datasheet is documented by using a template composed of

---

[1] https://hadoop.apache.org/.

[2] https://spark.apache.org/.

[3] Common services are services that will be part of every product derived from the SPL.

Table 1. Interactions defined for modeling variability

Interactions	JSON Property	Graphical Notation
Mandatory variation point	$MandatoryVP$	——————
Optional variation point	$OptionalVP$	------------
Alternative variation point	$AlternativeVP$	
Variant variation point	$VariantVP$	
Use	use	◆——————▶
Requires	$requires$	——————▶
Excludes	$excludes$	——/▶
Global variation point	$GlobalVariationPoint$	GVP
Specific variation point	$SpecificGlobalVariationPoint$	SVP

six items containing an *identification*, such as a number or code; a *textual name*, describing the main function; *the domain* in which this functionality is included; the *list of services* required for fulfilling this functionality; a *graphical notation*, which is a set of graphical artifacts showing the service interactions (as common and variant services); and a *JSON*[4] *file* specifying the services and their interactions. Within the *graphical notation item* (of the datasheets) any graphical artifact might be used; and particularly, we use an artifact based on variability annotation of collaboration diagrams (of UML). The required variability, according to the functionality to be represented, is attached to the diagrams by using the Orthogonal Variability Model (OVM) notation.

The complete set of interactions of our variability models is specified in Table 1. As we can see, they are divided into *variability types*, for denoting the variant interactions among services; *dependencies*, for denoting interactions between services; and *scope*, for specifying the scope of each variant point. There are two possibilities here: (1) a *Global Variation Point* specifies that if the variation point is instantiated in a specific way, it will be applied in that way for all functionality, including that variation point; and (2) a *Specific Variation Point* specifies that the instantiation of the variation point is particular for each functionality, including that variation point.

Examples of datasheets can be found in [2,3] for the marine ecology domain, and in [4] for the paleontology one. For instance, Table 2 shows the datasheet *Add New Excavations*, a variability example extracted from [4] to model processing variations of the *Load Excavation Data* service, which are identified by three possibilities (GPS, file, form).

---

[4] https://www.json.org/json-es.html.

Table 2. An example of the functional datasheet *Add New Excavations* (extracted from [4])

Id	FD2
Name	Add New Excavations
Domain	Palentology subdomain
Services	MMS-FA1.6, HI-LM1.31, PS-T4.20, HI-SLD2.3, ...
Variability Models	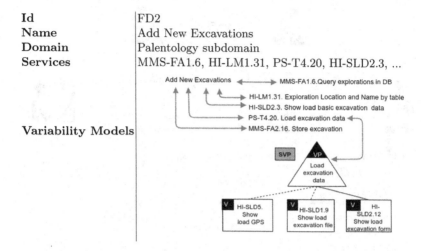

3 Our Approach in a Nutshell: Identifying Variety

Our answer to *RQ* is built by considering variety affecting a data processing pipeline, as shown in Fig. 1. Due to our focus on BDS, this pipeline indicates the essential activities of data analytics [6]: (1) *Data Ingestion*, which consists of extracting data from different sources, perhaps requiring some filtering and loading to adequate data; (2) *Data Preparation*, which consists of structuring, cleaning, transforming, and eventually integrating data before their processing; and (3) *Data Analysis*, which consists of transforming data into knowledge through different kinds of analyses (descriptive, predictive and/or prescriptive). Data visualization, the access point to results of this process, is split into two possibilities according to the working perspective of our proposal - top down and/or bottom up - as shown in Fig. 1. Let us introduce these cases.

Top-down Perspective: In this perspective, left hand side of Fig. 1, (1) given a domain problem, an expert user elaborates one or more hypotheses that should be tested through data analysis (i.e. Are data supporting this?); (2) then, the hypotheses are taken by data analysts who proceed to work into the data processing pipeline (3); finally, results are returned to verify the hypotheses (4) possibly visualizing data in different ways, and allowing hypothesis reformulation or process ending, alternatively (5).

Bottom-up Perspective: In this perspective, right hand side of Fig. 1, (1) given a domain problem, an expert user decides to launch an exploratory study to find out what data can reveal for this problem (i.e. What do data say?); (2) then, the study is carried out by data analysts (2), again by applying the data processing pipeline (3); and finally, results (findings) are returned to be validated with experts (4), alternately ending the process or reformulating the search (5).

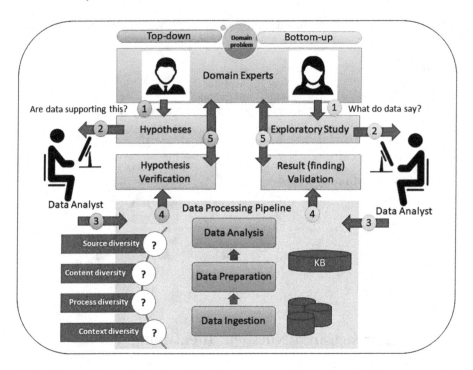

Fig. 1. Working perspectives identifying variety in a data processing pipeline

Expert users actively participate in the two last steps (4) and (5); however, they are also acting as domain consultants during the whole process. Of course, a zig-zag perspective combining top-down and bottom-up views would be valid too. As both perspectives might iterate according to reformulations, it would be perfectly possible to decide one way or the other, depending on particular results.

Until now, it seems that the data analysts' work is similar to ordinary cases of data science applications. However, during the data processing pipeline something else is happening. Variety is influencing decisions and guiding these processes, so each activity and/or result might be reused. As we can see from Fig. 1, variety is influencing design decisions by considering different diversities (source, content, process, and context). For instance, *source diversity* during *data ingestion* will help detect different data structures, acquisition techniques, etc.; meanwhile, *content diversity* will be focused on the way data should be transformed according to the business goals to be achieved, mostly considering source variations during *data preparation*. On the other hand, *process diversity* will help detect variations in data analysis techniques; and finally, *context diversity* will allow identify domain variations that may constrain or affect the results of the analysis during the whole pipeline.

All these variations will be stored in a knowledge base (KB in Fig. 1), so development/design decisions can be reused in similar situations, but possibly in different contexts. We store variations as datasheets, as were introduced in the previous section. For instance, Fig. 2 shows a datasheet to detect *data analysis diversity*. The first functionality, *Search for suitable model* comprises the actions needed to define the analysis goal, and then recovering the most suitable analysis technique for this goal. In this case, we can see the variability model associated to the variation point *analysis techniques*, which represents a case of alternative variability; that is, only one of the variants can be instantiated (for example, as *neural network, clustering with k-means*, etc.).

Once the most suitable technique is selected, another datasheet models its corresponding functionality. In Fig. 2, we can see the case for *neural network*, with two different models available from the KB. These models should be analyzed to determine their reusability for the present domain problem. Looking at the datasheet, we can see that there are two possibilities for variability models: (1) optional variation points, when the proposal and existing models show similarities; and (2) alternative variation points, when a new model should be created and stored[5], or it is possible to reuse an existing one.

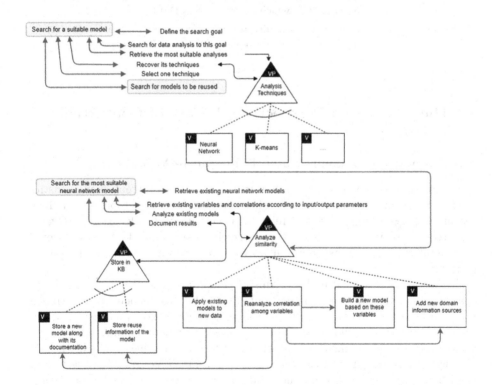

Fig. 2. Datasheets for modeling a case of data analysis diversity

---

[5] The <<require>> restriction implies that every new model is stored in the KB.

Table 3. Some water quality parameters (extracted from [8])

Parameter	Description
pH concentration	A measure of the relative amount of free hydrogen and hidroxyl ions in the water, used for testing acidity.
DO (Dissolved Oxygen)	A measure of oxygen dissolved in the water. A deficiency in DO is a sign of unhealthy river.
TDS (Total Dissolved Solids)	Any minerals, salts, metals dissolved in water. Provides a qualitative measure of the amount of dissolved ions present in the water.
EC (Electrical Conducivity)	Total amount of dissolved ions in the water which is a measure of salinity that affects the taste of potable water.
WT (Water Temperature)	Temperature affects the ability of water to hold oxygen, as well as the ability of organisms to resist certain pollutants. Aquatic organisms are dependent on certain temperature ranges for optimal health.
Turbidity	Turbidity results from suspended solids in the water, influences light penetration and makes the water cloudy or opaque. Such particles absorb heat in the sunlight, thus raising WT, which in turn lowers DO levels.
Nitrate	Excess levels of nitrates in water can create conditions that make it difficult for aquatic insects or fish to survive.
...	

4   The Bottom-up Perspective: A Case of Contextual Variety

4.1   Case Study Domain: Water Temperature Variation

The analysis of water resources refers to the study of inherent characteristics of their chemical, physical and biological structures; but also focuses on these resources as geographical objects related to each other. The location of a water resource is critical, since it can be near large cities, surrounded by roads, covered by vegetation, etc. - all aspects that a hydric resource monitoring system must consider.

Water quality is measured in terms of changes on chemical, ecological and spatial parameters, by analyzing not only figures but also dependencies. Among these parameters, we can find *pH concentration, DO (Dissolved Oxygen), water temperature*, etc., as shown in Table 3. The environment undoubtedly affects these parameters; for instance the nearby presence of mines impacts the pH of a water resource. It is clear then that the study of geographic objects and conditions of the communities where the water resource is located is important to determine its quality [8].

Figure 3 shows our proposal instantiated in a motivating case for the bottom-up perspective, where Mary, a domain expert, is concerned with detecting causes

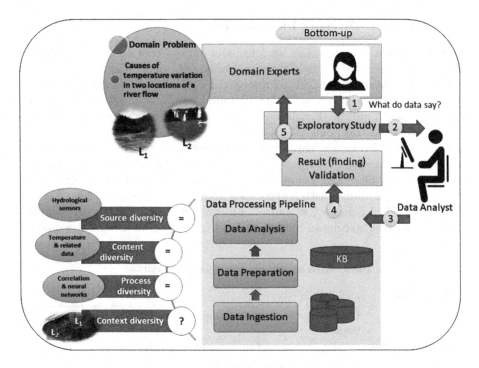

Fig. 3. Bottom-up perspective instantiated

of water temperature variation in two different locations of a river flow. Then, an exploratory study is conducted to investigate these possible causes; that is, the study explores a case of *contextual diversity* trying to answer how environmental/domain factors may affect design decisions in big data system development.

In order to avoid influences from the other types of diversity, we define a case where sources, content, and processing remain stable. That is, we collect data from the same sources and with the same format for two locations of the rivers (locators L1 and L2); transform and prepare data in the same way for L1 and L2; and use the same analysis techniques in both cases. In this way, we expect to reduce effects of additional variations on the case study. Therefore, we show here a concrete answer to RQ, where contextual diversity will be identified by relating inferences from a particular data set (bottom-up perspective). In other words, we will identify common and variable features of two locations (L1 and L2) trying to characterize both cases, and store this characterization along with their data analysis techniques *for* future reuse. It means that, in the future, given a hypothetical location L3, we might think of developing a Big Data system by reusing knowledge about techniques that were more suitable to this location and according to similarity of its characterization (development *with* reuse).

4.2 Contextual Diversity Identification

Data Ingestion. For our case, we selected a dataset containing water quality samples from lakes and rivers in the region of King County, Washington, United States[6]. This dataset has 1.589.362 rows and 25 columns representing variables, whose description is shown in Table 4. We used the same dataset for analyzing both locations L1 and L2 (Fig. 3), ensuring stability of the source and, therefore, with no variety at this point.

Implementation Decisions. We found that services provided by Apache Flume[7] were adequate for the batch processing (all at once) of this data. In addition, since we should filter and group data by parameter type under a bottom-up perspective (exploratory, with no guidance from user queries), we chose Apache Hbase[8] instead of Apache Cassandra[9]. Filtering should be mostly by column, given the multivalued column "parameter", where precisely a list of water quality

Table 4. Some relevant variables in the dataset

Column name	Description
Sample ID	Unique identifier to link samples collected from the same bottle.
Collect date time	Combined date and time sample was collected.
Depth (m)	Depth sample was collected (meters).
Site Type	General description of where sample was collected (e.g., marine beach, large lake, river, stream).
Area	Description of general area sample was collected.
Locator	Coordinate information of where the sample was collected.
Site	Description of sample collection location.
Parameter	What was analyzed, typically multiple parameters per sample (Temperature, Dissolved oxygen, Field pH, Field Conductivity, Field ammonia nitrogen, etc.)
Value	Measure for this parameter.
Units	Units associated with the value (mg/L, deg C, pH, etc.)
Quality	Code indicating overall quality of the data: (0) - Quality unknown; (1) Good Data, Passes data manager QC., (2) - Provisional data, Limited QC., (3) Questionable/Suspect, (4) Poor/Bad data, (5) Value changed (see Steward Note). (6) Estimated value, (7) Missing value.
...	...

---

[6] https://data.kingcounty.gov/Environment-Waste-Management/Water-Quality/
vwmt-pvjw.

[7] https://flume.apache.org/.

[8] https://hbase.apache.org/.

[9] https://cassandra.apache.org/.

parameters is stored. As a final decision, we selected the *Site Type* corresponding to *rivers* to instantiate the case.

Data Preparation. First of all, we analyzed and filtered data to keep those relevant to our domain problem. The bottom-up perspective implied that there was no previous indication of candidate locations to be analyzed; so, we carefully looked at this dataset. For instance, we counted the number of occurrences for each parameter, realizing that three of them (*Total Coliform, Total Hydrolyzable Phosphorus, Salininty*) were really low. In a similar way, we counted data of rivers by area founding that *Green* and *Issaquah* grouped the greatest number of occurrences, and selected them as candidates for determining locations L1 and L2. Finally, as each area may have more than one locator (in this case, 4 for each one), we counted occurrences for these eight cases and also checked their number of parameters. For instance, Fig. 4 shows proportionaly the number of occurrences for each locator in *Green* river, where the locator 3106 was selected as L1. Similarly, we selected the locator 0631 in *Issaquah* as L2.

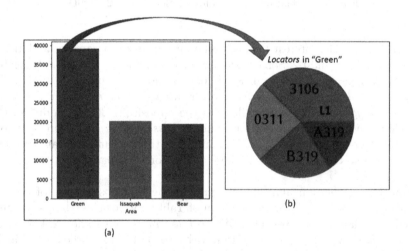

Fig. 4. Occurrences for each locator in Green river - 3106 (L1)

Once we had determined L1 and L2, we proceeded to transform the structure of parameters from rows to columns, since the following analyses would need one column for each parameter instead of a row with a list of them[10].

Following, in order to assess relationships between the different variables and our target (*Water Temperature*), we performed a Pearson correlation analysis on L1, detecting that the most related variables were *Total Nitrogen, Total Alkalinity, Dissolved Oxygen Field, Dissolved Oxygen,* and *Conductivity Field* (left-hand

---

[10] To do so, we used the pivot operation of Pandas (https://pandas.pydata.org/docs/index.html).

side of Fig. 5). Then, we calculated correlations for the most significant parameters of L1, but with data of L2. In this case, as we observed several differences, we decided to replicate the analysis to detect the most significant parameters for L2 (right-hand side of Fig. 5)[11].

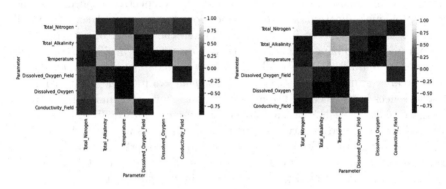

Fig. 5. Correlation analyses for L1 (left) and L2 (right)

Table 5 shows the intensity of each relation between a parameter and *Water Tempertaure* in L1 and L2. Some variables, such as *pH Field*, are strongly related to *Water Temperature* in L1; meanwhile, it is not the same case in L2. For instance, *Dissolved Oxygen Field* is more related to *Water Tempertaure* than *Dissolved Oxygen* in L2; meanwhile it is the opposite in L1. In addition, the intensity of the relations varies; for example *Dissolved Oxygen* decreases from -0.928462 (L1) to -0.703626 (L2). Summing up, the two intermediate columns of Table 5 show the differences of these values for each parameter ordered by incidence.

From the correlation analysis, we glimpsed the first possibilities of contextual diversity. Looking at the last column of Table 5, we can see the parameters ordered by variation with respect to *Water Temperature* in L1 and L2. For example, *pH Field* shows the largest variation with 0.011134 in L1, and 0.557562 in L2 (0,546428); meanwhile *Conductivity Field* shows the least variation (0,011824). As we can see, we applied the same transformations for both sites (L1 and L2), so content diversity remains the same.

Data Analysis. Figure 6 shows the datasheet instantiation for this case. Considering that there is no previous knowledge about this problem stored in our KB, we decided to analyze data through neural network modeling for estimating *Water Temperature* in L1 and L2 (*Select one technique <-> Neural Network* in

---

[11] In order to refine the correlation analysis, the transformed dataset was treated to mitigate influences of null values. Here, we tried several replacement alternatives (zero, mean value, etc.), to choose the better one.

Table 5. Correlations and variations among parameters and *Water Temperature*

Parameter	L1	L2	Ordered by variantion	Degree
Dissolved Oxygen	−0.928462	−0.703626	pH Field	0,546428
Dissolved Oxygen Field	−0.857255	−0.74100	Silica	0,454734
Total Alkalinity	0.591276	0.654430	Nitrite + Nitrate Nitrogen	0,376455
Conductivity Field	0.582982	0.594806	pH	0,374245
Total Nitrogen	−0.420579	−0.439275	Conductivity	0,301247
Orthophosphate Phosphorus	0.320313	0.110690	Ammonia Nitrogen	0,237398
Ammonia Nitrogen	0.257200	0.019802	Total Phosphorus	0,230068
Nitrite + Nitrate Nitrogen	−0.246479	−0.622934	Dissolved Oxygen	0,224836
Total Phosphorus	0.242684	−0.012616	Orthophosphate Phosphorus	0,209623
Conductivity	0.208746	0.509993	Total Suspended Solids	0,164463
Total Suspended Solids	−0.202858	−0.038395	Fecal Coliform	0,158106
Fecal Coliform	0.047256	0.205362	Turbidity	0,142344
Turbidity	−0.171750	−0.029406	E. coli	0,141807
E. coli	0.111787	0.253594	Dissolved Oxygen Field	0,116255
Silica	0.073081	0.527815	Total Alkalinity	0,063154
pH	−0.050635	0.424880	Enterococcus	0,050563
Enterococcus	0.042639	0.093202	Total Nitrogen	0,018696
pH Field	0.011134	0.557562	Conductivity Field	0,011824

Fig. 6). Then, our first model, M_1, was trained in L1[12]; and tested by checking variations between measured and estimated values (*Document Results <-> Store a new model along with its documentation* in Fig. 6). In this case, the average variation (calculated for a series of random samples) was **2.56841430**.

This first model M_1 was reused in L2 with no modifications in configuration or context (*Retrieve existing neural network models* in Fig. 6), resulting in **3.43957511** for average variation (Fig. 7 (a)). This fact led us to wonder whether considering contextual diversity would help decrease this figure. To do so, and thinking of correlation variations in Table 5 (*Retrieve existing variables and correlations according to input/output parameters* in Fig. 6), we defined a new model , M_2, with the same architecture as M_1 (no process variation) but changing input parameters. That is, meanwhile M_1 used *Total Nitrogen, Total Alkalinity, Dissolved Oxygen Field, Dissolved Oxygen* and *Conductivity Field*; in M_2 we changed by *Total Alkalinity, Dissolved Oxygen Field, Dissolved Oxygen, Nitrite + Nitrate Nitrogen, Conductivity Field, pH Field* and *Silica*. This new model was adjusted by contextual features in L2, so we would expect a better performance than in the case of M_1 (*Analyze existing models* in Fig. 6). Effectively, as Fig. 7 (b) shows, the average variation decreased to **2.303304169**.

Now, this new model might be retrieved for similar contexts in the future (*Document Results* in Fig. 6).

---

[12] We used Keras (https://keras.io/).

Result (finding) Validation. Our exploratory study shown that there are contextual differences that deserve more investigation. For instance, Fig. 8 shows both locations, L1 and L2, on a map. It is interesting to see that in spite of they belong to the same river, their environmental conditions are quite different: meanwhile L1 is surrounded by an urban area, L2 is in the middle of a dense forest. This glance at the map shows that these conditions may explain variations in water quality parameters; for instance, *pH Field* is higher in L2 - probably due to the forested area and the type of vegetation. Therefore, Mary might investigate additional data sources related to water uses in L1 and type of forest in L2.

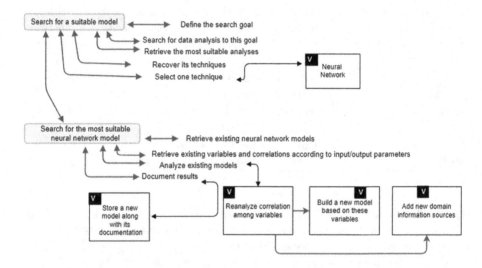

Fig. 6. Variability model (datasheet) instantiated

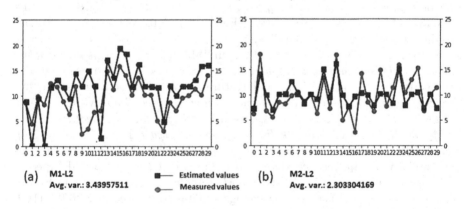

Fig. 7. Average variation for M_1 and M_2 in L2

Fig. 8. L1 and L2 in the geographic space

The bottom-up perspective may be influenced by the dataset itself, since it is the starting point of the process. Therefore, we checked the variable types, but also to the number of records and even decisions for preparing data. For instance, Table 5 shows that *pH Field* and *Silica* impact on *Water Temperature* differently when comparing L1 and L2. This fact may occur due diverse causes: (1) L1 could record fewer instances of these parameters than L2, so influence on correlation is negligible; or, on the contrary, (2) L1 could record many more instances than L2 determining that relations to *Water Temperature* are weak. Even when these two possibilities were controlled, it is still possible that the values of the parameters range differently for the two geographic areas, and consequently affect *Water Temperature*. This last possibility would be easily identified by Mary when analyzing findings of the process.

5   Conclusions

In this paper, we introduced a proposal for identifying and modeling variety to build reusable Big Data systems. Particularly, we discussed the case of contextual variety by following a case study as a motivating example, where domain expert participation was omitted in practice. However, as this participation is a core aspect, our work is currently extended by applying both perspectives of the proposal on the analysis of groundwater variation depending on contextual diversity in different geographic areas. This work is performed in collaboration with domain experts of the National Institute of Agriculture Technology[13], Patagonia, Argentina.

---

[13] https://inta.gob.ar/altovalle.

References

1. Borrison, R., Klöpper, B., Chioua, M., Dix, M., Sprick, B.: Reusable big data system for industrial data mining - a case study on anomaly detection in chemical plants. In: Yin, H., Camacho, D., Novais, P., Tallón-Ballesteros, A.J. (eds.) IDEAL 2018. LNCS, vol. 11314, pp. 611–622. Springer, Cham (2018). https://doi.org/10.1007/978-3-030-03493-1_64
2. Buccella, A., Cechich, A., Arias, M., Pol'la, M., Doldan, S., Morsan, E.: Towards systematic software reuse of GIS: Insights from a case study. Comput. Geosci. **54**, 9–20 (2013)
3. Buccella, A., Cechich, A., Pol'la, M., Arias, M., Doldan, S., Morsan, E.: Marine ecology service reuse through taxonomy-oriented SPL development. Comput. Geosci. **73**, 108–121 (2014)
4. Buccella, A., Cechich, A., Porfiri, J., Diniz Dos Santos, D.: Taxonomy-oriented domain analysis of GIS: a case study for paleontological software systems. ISPRS Int. J. Geo-Inf. **8**(6), 270 (2019). https://www.mdpi.com/2220-9964/8/6/270
5. Custers, B., Uršič, H.: Big data and data reuse: a taxonomy of data reuse for balancing big data benefits and personal data protection. Int. Data Priv. Law **6**(1), 4–15 (2016)
6. Davoudian, A., Liu, M.: Big data systems: a software engineering perspective. ACM Comput. Surv. **53**(5), 1–39 (2020)
7. Klein, J.: Reference architectures for big data systems, Carnegie Mellon University's software engineering institute blog (2017). http://insights.sei.cmu.edu/blog/reference-architectures-for-big-data-systems/. Accessed 9 Jun 2021
8. Loucks, D.P., van Beek, E.: Water Resource Systems Planning and Management. Springer, Cham (2017). https://doi.org/10.1007/978-3-319-44234-1
9. Najafabadi, M.M., Villanustre, F., Khoshgoftaar, T.M., Seliya, N., Wald, R., Muharemagic, E.: Deep learning applications and challenges in big data analytics. J. Big Data **2**(1), 1–21 (2015). https://doi.org/10.1186/s40537-014-0007-7
10. Osycka, L., Buccella, A., Cechich, A.: Identificación de variedad contextual en modelado de sistemas big data. In: Memorias del XXVII Congreso Argentino de Ciencias de la Computación (CACIC), pp. 367–376. Red de Universidades con Carreras en Informática (2021)
11. Pasquetto, I., Randles, B., Borgman, C.: On the reuse of scientific data. Data Sci. J. **16**(8) (2017)
12. Xie, Z., Chen, Y., Speer, J., Walters, T., Tarazaga, P.A., Kasarda, M.: Towards use and reuse driven big data management. In: Proceedings of the 15th ACM/IEEE-CS Joint Conference on Digital Libraries, pp. 65–74. Association for Computing Machinery (2015)

Best Practices for Requirements Validation Process

Sonia R. Santana[1]([✉]) [iD], Leandro R. Antonelli[2] [iD], and Pablo J. Thomas[3] [iD]

[1] Facultad de Ciencias de la Administración, Universidad Nacional de Entre Ríos, Concordia,
Entre Ríos, Argentina
sonia.santana@uner.edu.ar
[2] Laboratorio de Investigación y Formación en Informática Avanzada (LIFIA),
Facultad de Informática, Universidad Nacional de La Plata, Buenos Aires, Argentina
leandro.antonelli@lifia.info.unlp.edu.ar
[3] Instituto de Investigación en Informática (LIDI), Facultad de Informática,
Universidad Nacional de La Plata, Buenos Aires, Argentina
pthomas@lidi.info.unlp.edu.ar

Abstract. One of the most important phases in software development projects is the validation of requirements. Erroneous requirements, if not detected on time, can cause problems, such as additional costs, failure to meet expected objectives and delays in delivery. For these reasons, it is beneficial to invest efforts to this task. This paper aims to identify best practices that can help to carry out the Requirements Validation process. The best practices are determined by the analysis of software requirements validation approaches proposed in recent years, in order to evaluate their characteristics with the "Way-of" framework and the reference model for technical reviews.

Keywords: Requirements Engineering · Requirements validation ·
Requirements validation best practices

1 Introduction

In the framework of Requirements Engineering (RE), requirements validation is a fundamental task in any software engineering project and should be a continuous process in the system development life cycle. The main objective of requirements validation is to confirm that the requirements really describes the user needs and expectations [1–3] and that they are complete, correct and consistent [4] among other characteristics.

According to Kotonya [5] requirements validation refers to verifying the consistency, completeness and correctness of the requirements document [3], and also states that requirements must be: valid, understandable, consistent, traceable, complete, true and verifiable. According to Bahill [6] the requirements validation process consists firstly in ensuring that a set of requirements are: correct, complete and consistent; secondly that a model can be created and meets the requirements; and finally, that a software solution can be built and tested in the real world to demonstrate that it meets the requirements of the stakeholders.

P. Pesado and G. Gil (Eds.): CACIC 2021, CCIS 1584, pp. 139–156, 2022.
https://doi.org/10.1007/978-3-031-05903-2_10

Working on requirements validation is becoming a challenge for teams, customers and users. There are different causes that impose problems of communication, control, knowledge sharing, trust and delays in software development [7].

Based on the results obtained in [8], this work seeks to identify good practices that can help in the development of requirements validation approaches. The identification and characterization of good practices is performed through the analysis of software requirements validation approaches by means of a literature review and an evaluation of various requirements validation approaches.

This paper is organized as follows. Section 2 describes a literature review performed to identify the different approaches to practices related to requirements validation. Then, Sect. 3 analyzes the characteristics of the approaches with the "Way-of" framework and the reference model for technical reviews, in order to know their characteristics, information needs and constraints. Section 4 presents the results of the comparative analysis of the selected approaches, where the contributions of the works for the characterization of good practices for the development of new approaches in the Requirements Validation process were identified. Finally, conclusions and future work are presented.

2 Literature Review

This section describes a three-phase process: search, selection and evaluation of articles for a literature review of requirements validation. To start the literature review, the following question was posed: What are the main practices related to the requirements validation process? The objective of this research question was to identify the main practices related to the requirements validation process.

2.1 Search for Articles

The articles were searched in the following sources: IEEE, Elsevier, Springer and ACM Digital Library. Articles published in Spanish and English in recent years were considered. The text used for the queries to search for papers that could answer to the initial question, are the following:

- ("requirements") and ("validation" or "validate" or "validity") and ("methodology" or "technique" or "method" or "tool")
- ("requirements" or "requirements") and ("validation" or "validate" or "validity") and ("methodology" or "technique" or "method" or "tool")

2.2 Item Selection

For the selection process, several criteria for inclusion and exclusion of articles were established:

- The selected articles are directly related to the topic of requirements validation in the area of RE.

- Articles discussing requirements validation in the testing phase of the developed software or in the implementation under test with respect to the requirements were not considered.
- Regarding articles using terms such as "Systematic literature review", "Literature review", "Systematic analysis", only those proposing new approaches to requirements validation were considered.

2.3   Evaluation of Articles

For the evaluation of the articles, a list of criteria was developed and adapted from [9], which allowed the final articles to be selected based on four criteria listed in Table 1.

Table 1. Evaluation criteria.

Section	Criteria
Introduction	1. Does the introduction provide an overview of the input for requirements validation? 2. Is the purpose/objective of the research clearly defined?
Methodology	3. Is the research methodology clearly defined?
Results	4. Are the findings clearly stated, and do the results help to solve requirements validation problems?
Conclusion	5. Are there any limits or restrictions imposed on the conclusion statement?

The initial search provides 30767 articles, after eliminating duplicated articles and applying the inclusion and exclusion criteria defined in Sect. 2.2, 7898 articles were preselected. After reading the titles, abstracts and conclusions, 486 relevant articles were preselected. Finally, after applying the evaluation criteria defined in Table 1, 38 papers were selected.

For each of the articles, the year of publication, the name and domain of the application were identified. The results are presented in Table 2.

Table 2. Selected articles from the literature review.

Ref. Bibliog	Year	Name	Application domain
[10]	2007	FBCM	Company
[11]	2007	CPN Tool	Health care
[12]	2008	AutoPA3.0	Library system
[13]	2009	EuRailCheck	Transportation (train)

(*continued*)

Table 2. (*continued*)

Ref. Bibliog	Year	Name	Application domain
[14]	2009	ACE Framework	Online discussion/forum
[15]	2009	–	Safety critical
[16]	2010	SQ$^{(2)}$	Production line
[17]	2010	–	Generic domain document
[18]	2011	CoReVDO	Distributed systems
[19]	2011	MaramaAI	ATM System
[20]	2011	–	Vehicle electronics
[21]	2011	VRP	Embedded software
[22]	2012	–	Library system
[23]	2012	Othello	Transportation
[24]	2013		Health system
[25]	2013	–	Multi-agent system
[26]	2014	CuRV	Smart Phone
[27]	2014	SpecQua	Health care system
[28]	2014	–	Company
[29]	2015	ReVAMP	Company
[30]	2015	–	Company shopping on line
[31]	2015	SimTree	Sanitary device
[32]	2016	–	Transportation
[33]	2016	MobiMEReq	Mobile applications
[34]	2016	ASATF	Search system
[35]	2016	TestMEReq	Company
[36]	2017	–	Railway signaling
[37]	2017	SHPbench	Automobile driver assistance
[38]	2018	–	Nanosatellite and satellite control system
[39]	2018	–	Process control system
[40]	2018	–	Driver assistance system
[41]	2019	–	Educational Institution
[42]	2019	RM2PT	Company
[43]	2019	–	Course management system
[44]	2020	ValFAR	Course management system
[45]	2020	–	Automotive industry
[46]	2020	–	Civil aviation systems
[47]	2021	–	Train control system

This literature review made it possible to pre select 38 articles from which four requirements validation approaches were subsequently selected that detail a methodological guide for their mode of thinking, modeling, concepts, management and work. Table 3 shows the result of the final selection.

Table 3. Selected approaches to requirements validation.

Ref. Bibliog	Ref. Study	Year	Name	Application domain
[10]	E1	2007	FBCM	Company
[15]	E2	2009	From informal requirements to property-driven formal validation	Safety-critical system
[18]	E3	2011	CoReVDO	Distributed systems
[46]	E4	2020	A methodology of requirements validation for aviation system development	Civil aviation systems

3   Evaluation of Approaches

The selected approaches were evaluated using the "Way-of" framework and the reference model for technical reviews to understand their characteristics, information needs and constraints.

3.1   Definition of the "Way-of" Framework for the Evaluation of Approaches

There are approaches that introduce models with little knowledge, others offer algorithms, or at least explicit procedures to build a specific model or verify it. Yet other approaches provide informal but practical suggestions for obtaining a model. It is therefore advisable to distinguish between the "way of modeling" and the "way of working" of an approach. The "way of control" is essentially related to time, cost and quality control of the information systems development process and its products.

According to Seligmann [48], approaches can be described by differentiating between a way of modeling, a way of working and a way of controlling. However, to really understand an approach Sol [49] and Kensing [50] consider necessary to know its underlying philosophy or "way of thinking" used to look at organizations and information systems. The "Way-of" framework is used to evaluate approaches with five different aspects, each described with explanations of their contributions to the flexible management of the development process [51]. The five aspects of the "Way-of" framework are:

- **Way of Thinking:** Defines guidelines for the development of the approach, thus providing a perspective of the problem domain and making explicit the assumptions, principles and strategies about it.

144   S. R. Santana et al.

- **Modeling Form:** Provides concept information for modeling. Provides formalism and notation for expressing process models of the approach.
- **Way of Working:** Defines the structure of the process of the approach, the activities, tasks and the sequence in which they should be carried out.
- **How to Control:** Defines the means provided by the approach to determine how the approach is to be controlled and evaluated.
- **Form of Support**: Refers to the techniques, tools and/or work aids that support the execution of the process of the approach.

Based on these aspects, a set of questions were developed to evaluate them. These questions are presented in the Table 4.

Table 4. Set of questions for assessing aspects of the "Way of" framework

Form of	Evaluation questions
Think (Framing the approach)	• What is the function of focus? • What are the components of the approach? • What is the focus environment? • What are the components of the focus environment? • What are the characteristics of the approach and its environment?
Model (Operational process oriented to the product of the approach)	• What model does the approach use? • Does the approach describe the components and their relationships within the models used? • Does the approach describe the relationships between the models used?
Work (Operational process oriented to the approach process)	• What are the activities and tasks of the approach? • What are the responsibilities of the focus activities and tasks?
Control (Process oriented to control and evaluate the approach)	• What is the form of control used by the approach? • What are the objectives measured through indicators?
Support/Support (Process to support implementation of the approach)	• What are the techniques, tools and/or job aids that support the approach? • What standards does the approach use?

3.2 Evaluation of Approaches According to the "Way-of" Framework

This subsection describes the analysis of each of the selected approaches listed in Table 2. The analysis consists in answering each one of the questions described in Table 4. The answers for each approach are presented in Tables 5, 6, 7 and 8.

Table 5. Evaluation of the E1 approach

Form of	E1 approach
Think	**Function**: define and validate business requirements that are used as software functional requirements **Components**: collaborative fact models **Environment**: business systems **Environment components**: system planning, requirements analysis and specification
Think	**Characteristics:** to assess the integrity of the organization's fundamental goals and objectives for the development of the information technology system
Model	**Model**: BSC (Balanced score card) is used to develop strategies in companies **Components**: Perspective: viewpoints of the strategic objective- Critical processes - Strategic objectives of the company - Key performance indicator (KPI) - Cause-effect relationship between strategic objectives **Relationship between components**: Visualizing the BSC strategy, adding objectives by observation, analyzing the strategy structure, evaluating the validity of the strategy structure, and extracting system functions
Work	The approach proposes a 5-step vision: 1. Visualize the BSC strategy. Generate an objective analysis tree 2. Add targets per observation 3. Analyze the strategy structure. Assign KPIs to each strategic objective 4. Evaluate the validity of the strategy structure, by means of analyses statistical analysis of KPI data 5. Extract results and refine the target analysis tree
Control	**Control**: library of 700 KPIs for developers to choose KPIs easily divided into 4 perspectives: financial, customer, business process, learning and knowledge
Support	**Techniques**: objective analysis tree, field observation cards, strategy map and collaboration matrix

Table 6. Evaluation of the E2 approach

Form of	E2 approach
Think	**Function**: to validate a requirements specification written in informal language **Components**: formal models for requirements validation **Environment**: complex safety-critical systems **Components of the environment**: specification of requirements written in an informal language **Characteristics**: Formalize requirements in the use of the Unified Modeling Language (UML) and in the use of a Controlled Natural Language (CNL), based on a subset of the Property Specification Language (PSL)

(continued)

Table 6. (*continued*)

Form of	E2 approach
Model	**Model**: Unified Modeling Language (UML), Controlled Natural Language (CNL) and Property Specification Language (PSL) **Components**: class diagram, state machines and sequence diagrams **Relationship between the components**: First, the approach provides for an informal analysis of the requirements document to categorize each requirement. Then, each requirement fragment is formalized according to the categorization by means of UML diagrams and the use of a Controlled Natural Language. Finally, an automatic formal analysis is performed to identify possible flaws in the formalized requirements
Work	The approach proposes a 3-phase vision: 1. Phase of analysis of informal requirements based on inspections to identify failures
Work	2. Formalization phase. Each requirement fragment identified in the informal analysis phase specifying the corresponding UML concepts and diagrams and/or CNL constraints. Link the UML elements with the textual requirements 3. Formal validation phase. Verify, reduce and validate the fragments of formalized requirements
Control	**Control:** Vacuity check: check whether a given property is permanently maintained. Coverage check: check which elements of the considered formalized requirement fragment have been stimulated (covered) by a generated trace. Security analysis: identify the causes that lead to the violation of a property, i.e., identify variables of interest that are causes of a specific violation so that advanced algorithms can compile a description of the causes and organize them in the form of a fault tree
Support	**Techniques**: class diagram, sequence diagram, state machines and controlled natural language **Tools**: developed based on industry standards: IBM Rational RequisitePro (RRP), interfaced with Microsoft Word, and IBM Rational Software Architect (RSA), to support the informal analysis phase and the traceability of the link between informal requirement fragments and their counterparts

Table 7. Evaluation of the E3 approach

Form of	E3 approach
Think	**Function**: to verify, negotiate and validate distributed requirements through a set of activities **Components**: multiple viewpoints and cognitive competencies of stakeholders in a distributed and collaborative process **Environment**: distributed systems **Components of the environment**: requirements specification document **Characteristics**: work with geographically distributed teams and include the client in the collaborative validation process

(continued)

Table 7. (*continued*)

Form of	E3 approach
Model	**Model**: Unified Modeling Language (UML) **Components**: activity diagram **Relationship between components**: It features three main activities - Organization, Distributed Verification and Collaborative Validation
Work	The approach proposes a 3-phase vision: 1. Organizational. Responsibility: Analyst. RE Team 2. Verification of functional and Non-functional requirements Responsibility: SR Team 3. Collaborative validation. Responsibility: SR Team
Control	**Control**: rate of motivation and commitment among stakeholders. If the rate is positive, the agreement between the parties is strong; otherwise, it is moderate or negative Global consistency where a requirement must not contradict other requirements established after global requirements integration
Support	**Techniques**: checklist, point of view oriented, inspection, review and prototyping
Support	**Standard**: IEEE 830 for quality assurance of requirements and deliverables

Table 8. Evaluation of the E4 approach

Form of	E4 approach
Think	**Function**: validate a requirements specification and comply with ARP4754A certification, special consideration of the civil aircraft and systems development life cycle **Components**: formal models for requirements validation **Environment**: aviation systems **Components of the environment**: specification of requirements **Characteristics**: Validate requirements considering aircraft design features and certification regulations during the product development life cycle
Model	**Model**: validation process model based on compliance with the objects required by SAE ARP 4754A certification regulators. **Components**: Validation plan, validation rigor, correctness and correctness checks, validation matrix (initial and final) and validation report. **Relationship between components**: First, the approach develops a validation plan, then creates a validation matrix where it verifies correctness and checks completeness to perform its update. Finally, it generates a summary to review the validation activities
Work	The approach proposes a process divided into: 1. Create validation plan 2. Create a requirements validation matrix 3. Generate a requirements correction checklist 4. Generate a requirements completeness checklist 5. Update the requirements validation matrix 6. Requirements validation summary 7. Review requirements validation activities

(*continued*)

Table 8. (*continued*)

Form of	E4 approach
Control	**Control:** Checklists check the accuracy and completeness of the requirements
Support	Techniques: checklist, testing, inspection and review Tools: For requirements management, ReQtest, IBM rational doors, Visure requirements Standard: EIA 632 for the requirements validation process

3.3 Definition of the Reference Model to Perform a Technical Review to the Evaluation of the Approaches

This sections presents he reference model for technical reviews used by Pressman [52] to identify complementary processes to the "Way-of" framework in the evaluation of approaches [51].

This model considers the following activities: roles of individuals, planning and preparation, meeting structure, and correction and verification (see Fig. 1).

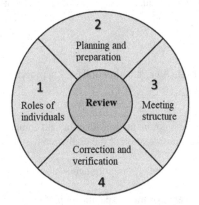

Fig. 1. Reference model for technical reviews [52].

Each of the characteristics of the reference model helps to define the level of formality of the review which is increased when: 1) distinct roles for reviewers are explicitly defined, 2) there is a sufficient amount of planning and preparation for the review, 3) a distinct structure for the review (including internal tasks and work products) is defined, and 4) follow-up by reviewers takes place for any corrections that are made [52].

Thus, three complementary activities can be defined for the requirements validation process:

- **Planning**. Activity that allows defining in advance the objectives to be obtained, the requirements to be reviewed, the people who will participate, the procedure, techniques and tools to be used. If these suggested elements are found, it can be called requirements validation plan.

- **Defect management.** It comprises the identification and documentation of defects as a result of the application of the validation process, but additionally the actions for their correction must be established, which finally leads to a follow-up and control of the defect status.
- **Acceptance.** The requirements acceptance is the last activity of the validation process; this activity is the point where the Requirements Engineering is connected with the next phase of the software development life cycle. This activity ensures the customer's acceptance of the specifications, conditions, constraints and quality parameters to be verified later in the software product.

3.4 Evaluation of Approaches According to the reference Model for Technical Reviews

Table 9 shows the evaluation of the approaches according to the complementary activities obtained from the reference model, in which the word YES indicates that the approach contributes to the requirements validation process, while the word NO indicates that the approach does not contribute to the requirements validation process.

Table 9. Evaluation of the approaches according to complementary processes of the reference model.

Activity	Approaches			
	E1	E2	E3	E4
Planning	NO	NO	YES	YES
Defects management	YES	YES	YES	YES
Acceptance	NO	NO	NO	YES

4 Comparative Analysis

From the analysis of the results obtained in the evaluation of the approaches indicated in the Table 9 and the features evaluated in the approaches in Tables 5, 6, 7, and 8, different characteristics are extracted that are qualified as evaluation points for the selected approaches. Table 10 lists the characteristics identified, where YES indicates that the approach contributes to the characteristic, while NO indicates that the approach does not contribute to the characteristic.

Table 10. Evaluation of requirements validation approaches.

Feature	Approaches			
	E1	E2	E3	E4
Definition				
Define the function	YES	YES	YES	YES
Defines the components	YES	YES	YES	YES
Defines the environment	YES	YES	YES	YES
Defines components of the environment	YES	YES	YES	YES
Define characteristics	YES	YES	YES	YES
Modeling				
Specifies the model	YES	YES	YES	YES
Specifies model components	YES	YES	YES	YES
Specifies relationships between model components	YES	YES	YES	YES
Operation				
Define a plan	NO	NO	YES	YES
Defines phases, activities and tasks	YES	YES	YES	NO
Define requirements to be validated	YES	YES	YES	YES
Define task roles	NO	NO	YES	YES
Define functions and responsibilities of the roles	NO	NO	YES	YES
Control				
Defines the form of control	YES	YES	YES	YES
Define performance indicators	YES	NO	NO	NO
Define objectives measured by indicators	YES	NO	NO	NO
Defects management				
Identifies defects	YES	YES	YES	YES
Document defects	YES	YES	YES	YES
Defect tracking	YES	YES	YES	YES
Acceptance				
Considers customer acceptance of specifications	NO	NO	YES	YES
Considers the customer's acceptance of the conditions	NO	NO	YES	YES
Consider the customer's acceptance of the restrictions	NO	NO	YES	YES
Consider the customer's acceptance of the quality parameters	NO	NO	YES	YES
Validates requirements in the software development life cycle	NO	NO	NO	YES
Support				

(continued)

Table 10. (*continued*)

Feature	Approaches			
	E1	E2	E3	E4
Define techniques	YES	YES	YES	YES
Define tools	NO	YES	NO	YES
Uses standards	NO	NO	YES	YES
Application domain				
Defines the application domain	YES	YES	YES	YES

The following results can be drawn from the comparative analysis of the selected approaches:

- In the process of definition, modeling, defect management and application domain definition, the approaches meet all the proposed characteristics.
- In the operation process, approaches E3 and E4 define a planning process for requirements validation. E1, E2, and E3 define a structure composed of phases, activities and tasks. While E4 only defines, a structure composed of activities and tasks. E2, E3 and E4 validate the requirements on the requirements specification and E1 validates the business requirements. E3 and E4 define roles and responsibilities.
- In the support process, approaches E1, E2, E3 and E4 define techniques for requirements validation. In addition, E2 and E4 specify tools. E3 implements the IEEE830 standard [53] for requirements specification and E4 the EIA632 [54] for the requirements validation process.
- In the control process, approaches E1, E2, E3 and E4 implement a form of control for the validation of requirements, but only E1 defines performance indicators and objectives based on these indicators.
- In the acceptance process, approaches E3 and E4 consider customer acceptance of specifications, constraints, conditions and quality parameters. E4 validates requirements throughout the software development life cycle.

From the analysis of requirements validation approaches, a classification and generalization of characteristics is proposed and detailed in Table 10. These characteristics are summarized in terms of best practice recommendations that can improve knowledge in the development of requirements validation approaches. They are described in Table 11.

Table 11. Description of good practices.

Feature	Good practices
Definition	Define the function of the focus Define the components of the approach Define the focus environment Define components of the focus environment Define characteristics of the approach
Modeling	Specify the model of the approach Specify components of the approach model Specify relationships between the components of the approach model
Operation	Define the planning approach Define phases, activities and tasks of the approach Define requirements that validate the approach Define roles of the focus tasks Define functions and responsibilities of the focus roles
Control	Define the form of focus control Define performance indicators for the approach Define objectives measured by focus indicators
Defects management	Define how to identify defects in the approach Define how to document defects in the approach Define how to track defects in the approach
Acceptance	Define customer acceptance of the approach specifications Define customer acceptance of the terms of the approach Define customer acceptance of the approach constraints Define customer acceptance of the quality parameters of the approach Define requirements validation in the software development life cycle of the approach
Support	Define approach techniques Define tools of the approach Define approach standards
Application domain	Define the application domain of the approach

5 Conclusions and Future Work

This work has focused on a fundamental process of Requirements Engineering: Validation. Thirty-eight papers were pre-selected and then focused on four of them. To answer the evaluation questions of the four selected approaches, the contributions of the papers were identified for the proposal of best practices for the Requirements Validation process.

All approaches were developed for specific application domains. While the approaches satisfy most of the characteristics of the "Way of" framework and the technical review reference model used for their evaluation, some issues that should be cope with were identified.

The approaches provide differences in the domain view to validate requirements with varying degrees of success. This success depends on the nature of the organization itself and the in-depth knowledge of the business to tailor the approach to the business and user needs.

There is little user/customer participation in the requirements validation process, integrated with a lack of vision on the part of management approaches on the level of responsibility, the degree of decision making and the balance between users and developers.

The approaches perform requirements validation on the requirements specification, i.e., they are not applied at different stages of the software development life cycle. While the E4 approach employs requirements validation throughout the system life cycle this is due to the use of aviation standard ARP4754A [20], where it supports certification of aircraft systems, addressing "the complete aircraft development cycle, from system requirements to system verification". Within the framework of the above observation, standards compliance is insufficient in addressing the implicit features expected in professional software development.

The low control and follow-up of the defects in the requirements through the use of performance indicators and the scarce definition of objectives to be evaluated by said indicators is evident.

The best practices proposed in Table 11 can help in the requirements validation process and solve the problems mentioned above.

As a future work, it is proposed to develop an approach based on the best practices presented above in order to improve the requirements validation process. The approach to be developed should first describe a framework for the development of the structure and then establish a model for the requirements validation process at the white box level in the internal structure of the design of the components of the system development. This allows to guarantee the quality of the product or system, focusing on the different stages of the software life cycle.

References

1. Laplante, P.A.: Requirements Engineering for Software and Systems. CRC Press, Boca Raton (2019)
2. Cheng, B.H.C., Atlee, J.M.: Current and future research directions in requirements engineering. In: Lyytinen, K., Loucopoulos, P., Mylopoulos, J., Robinson, B. (eds.) Design Requirements Engineering: A Ten-Year Perspective. LNBIP, vol. 14, pp. 11–43. Springer, Heidelberg (2009). https://doi.org/10.1007/978-3-540-92966-6_2
3. Kotonya, G., Sommerville, I.: Requirements Engineering: Processes and Techniques. Wiley, England (1998)
4. Pfleeger, S.L.: Software Engineering - Theory and Practice. Prentice Hall, Hoboken (1998)
5. Kotonya, G., Sommerville, I.: Requirements Engineering: Processes and Techniques. Wiley, England (2000)
6. Bahill, A.T., Henderson, S.J.: Requirements development, verification, and validation exhibited in famous failures. Syst. Eng. **8**, 1–14 (2005). https://doi.org/10.1002/sys.20017
7. Loucopoulos, P., Karakostas, V.: System Requirements Engineering, McGraw-Hill, London (1995). ISBN 0-07-707843-8

8. Santana, S.R., Antonelli, L., Thomas, P.: Evaluación de metodologías para la validación de requerimientos. In: XXVII Congreso Argentino de Ciencias de la Computación (CACIC), pp. 419–428 (2021). ISBN 978 -987-633-574-4
9. Aiza, M.N., Massila, K., Yusof, M.M., Sidek, S.: A review on requirements validation for software development. J. Theor. Appl. Inf. Technol. **96**(11) (2018)
10. Kokune, A., Mizuno, M., Kadoya, K., Yamamoto, S.: FBCM: strategy modeling method for the validation of software requirements. J. Syst. Softw. **80**(3), 314–327 (2007)
11. Machado, R.J., Lassen, K.B., Oliveira, S., Couto, M., Pinto, P.: Requirements validation: execution of UML Models with CPN tools. Int. J. Softw. Tools Technol. Transfer **9**(3–4), 353–369 (2007). https://doi.org/10.1007/s10009-007-0035-0
12. Li, D., Li, X., Liu, J., Liu, Z.: Validation of requirement models by automatic prototyping. Innov. Syst. Softw. Eng. **4**(3), 241–248 (2008)
13. Cavada, R., et al.: Tonetta: supporting requirements validation: the EuRailCheck tool. In: ASE2009 - 24th IEEE/ACM International Conference on Automated Software Engineering, pp. 665–667 (2009)
14. Jureta, I., Mylopoulos, J., Faulkner, S.: Analysis of multi-party agreement in requirements validation. In: 17th IEEE International Requirements Engineering Conference, pp. 57–66 (2009)
15. Cimatti, A., Roveri, M., Susi, A., Tonetta, S.: From informal requirements to property-driven formal validation. In: Cofer, D., Fantechi, A. (eds.) FMICS 2008. LNCS, vol. 5596, pp. 166–181. Springer, Heidelberg (2009). https://doi.org/10.1007/978-3-642-03240-0_15
16. Aceituna, D., Do, H., Lee, E.S.: SQ^(2): an approach to requirements validation with scenario question. In: Asia Pacific Software Engineering Conference, pp. 33–42 (2010)
17. Kof, L., Gacitua, R., Rouncefield, M., Sawyer, P.: Ontology and model alignment as a means for requirements validation. In: IEEE Fourth International Conference on Semantic Computing, pp. 46–51 (2010)
18. Sourour, M.D., Zarour, N.: A methodology of collaborative requirements validation in a cooperative environment. In: 10th International Symposium on Programming and Systems, Algiers, Algeria, pp. 140–147 (2011)
19. Kamalrudin, M., Grundy, J.: Generating essential user interface prototypes to validate requirements. In: 26th IEEE/ACM International Conference on Automated Software Engineering (ASE 2011), pp. 564–567 (2011)
20. Holtmann, J., Meyer, J., von Detten, M.: Automatic validation and correction of formalized, textual requirements. In: IEEE Fourth International Conference on Software Testing, Verification and Validation Workshops, pp. 486–495 (2011)
21. Aceituna, D., Do, H., Lee, S.W.: Interactive requirements validation for reactive systems through virtual requirements prototype. In: Model-Driven Requirements Engineering Workshop, MoDRE 2011, pp. 1–10 (2011)
22. Sharma, R., Biswas, K.K.: Using norm analysis patterns for automated requirements validation. In: Second IEEE International Workshop on Requirements Patterns (RePa), pp. 23–28 (2012)
23. Cimatti, A., Roveri, M., Susi, A., Tonetta, S.: Validation of requirements for hybrid systems. ACM Trans. Softw. Eng. Methodol **21**(4), 1–34 (2012)
24. Felderer, M., Beer, A.: Using defect taxonomies for requirements validation in industrial projects. In: 21st IEEE International Requirements Engineering Conference, RE 2013 - Proceedings, pp. 296–301 (2013)
25. Gaur, V., Soni, A.: A fuzzy traceability vector model for requirements validation. Int. J. Comput. Appl. Technol. **47**(2/3), 172–188 (2013)
26. Lee, Y.K., In, H.P., Kazman, R.: Customer requirements validation method based on mental models. In: 21st Asia-Pacific Software Engineering Conference, vol. 1, pp. 199–206 (2014)

27. Rodrigues, A.: Quality of requirements specifications - a framework for automatic validation of requirements. In: Proceedings of the 16th International Conference on Enterprise Information Systems, pp. 96–107 (2014)
28. Nazir, S.: A process improvement in requirement verification and validation using ontology. In: Asia-Pacific World Congress on Computer Science and Engineering, pp. 1–8 (2014)
29. Saito, S., Hagiwara, J., Yagasaki, T., Natsukawa, K.: ReVAMP: requirements validation approach using models and prototyping, practical cases of requirements engineering in end-user computing. J. Inf. Process. **23**(4), 411–419 (2015)
30. Ali, N., Lai, R.: A method of software requirements specification and validation for global software development. Requirements Eng. **22**(2), 191–214 (2015). https://doi.org/10.1007/s00766-015-0240-4
31. Zafar, S., Farooq-Khan, N., Ahmed, M.: Requirements simulation for early validation using behavior trees and Datalog. Inf. Softw. Technol. **61**, 52–70 (2015)
32. Miao, W., et al.: Automated requirements validation for atp software via specification review and testing. In: Ogata, K., Lawford, M., Liu, S. (eds.) ICFEM 2016. LNCS, vol. 10009, pp. 26–40. Springer, Cham (2016). https://doi.org/10.1007/978-3-319-47846-3_3
33. Yusop, N., Kamalrudin, M., Sidek, S., Grundy, J.: Automated support to capture and validate security requirements for mobile apps. Commun. Comput. Inf. Sci. **671**, 97–112 (2016)
34. El-Attar, M., Abdul-Ghani, H.A.: Using security robustness analysis for early-stage validation of functional security requirements. Requirements Eng. **21**(1), 1–27 (2014). https://doi.org/10.1007/s00766-014-0208-9
35. Moketar, A., Kamalrudin, M., Sidek, S., Robinson, M., Grundy, J.: An automated collaborative requirements engineering tool for better validation of requirements (2016)
36. Rosadini, B., Ferrari, A., Gori, G., Fantechi, A., Gnesi, S., Trotta, I., Bacherini, S.: Using NLP to detect requirements defects: an industrial experience in the railway domain. In: Grünbacher, P., Perini, A. (eds.) REFSQ 2017. LNCS, vol. 10153, pp. 344–360. Springer, Cham (2017). https://doi.org/10.1007/978-3-319-54045-0_24
37. Buchholz, C., Vorsatz, T., Kind, S., Stark, R.: SHPbench - a smart hybrid prototyping based environment for early testing, verification and (user based) validation of advanced driver assistant systems of cars. Procedia CIRP **60**, 139–144 (2017)
38. Stachtiari, E., Mavridou, A., Katsaros, P., Bliudze, S., Sifakis, J.: Early validation of system requirements and design through correctness-by-construction. J. Syst. Softw. **145**, 52–78 (2018)
39. Chahin, A., Paetzold, K.: Planning validation & verification steps according to the dependency of requirements and product architecture. In: IEEE International Conference on Engineering, Technology and Innovation (ICE/ITMC), pp. 1–6 (2018)
40. Gruber, K., Huemer, J., Zimmermann, A., Maschotta, R.: Automotive requirements validation and traceability analysis with AQL queries. In: IEEE International Systems Engineering Symposium (ISSE), pp. 1–7 (2018)
41. Bayona-Oré, S., Chamilco, J., Perez, D.: Software process improvement: requirements management, verification and validation. In: 14th Iberian Conference on Information Systems and Technologies (CISTI), pp. 1–5 (2019)
42. Yang, Y., Ke, W., Li, X.: RM2PT: requirements validation through automatic prototyping. In: IEEE 27th International Requirements Engineering Conference (RE), pp. 484–485 (2019)
43. Atoum, I.: A scalable operational framework for requirements validation using semantic and functional models. In: Proceedings of the 2nd International Conference on Software Engineering and Information Management (ICSIM 2019), New York, pp. 1–6. Association for Computing Machinery (2019)

44. Alshareef, S.F., Maatuk, A.M., Abdelaziz, T.M., Hagal, M.: Validation framework for aspectual requirements engineering (ValFAR). In: Proceedings of the 6th International Conference on Engineering & MIS (ICEMIS'20), Article 42, New York, pp. 1–7. Association for Computing Machinery (2020)
45. Iqbal, D., Abbas, A., Ali, M., Khan, M.U.S., Nawaz, R.: Requirement validation for embedded systems in automotive industry through modeling. IEEE **8**, 8697–8719 (2020)
46. Fei, X., Bin, C., Siming, Z.: A methodology of requirements validation for aviation system development. In: Chinese Control and Decision Conference (CCDC), pp. 4484–4489 (2020)
47. Mashkoor, A., Leuschel, M., Egyed, A.: Validation obligations: a novel approach to check compliance between requirements and their formal specification (2021)
48. Seligmann, P.S., Wijers, G.M., Sol, H.G.: Analyzing the structure of IS methodologies, an alternative approach. In: Proceedings of the First Dutch Conference on Information Systems, Amersfoort, The Netherlands (1989)
49. Sol, H.G.: A feature analysis of information systems design methodologies: methodological considerations. In: Olle, T.W., Sol, H.G., Tully, C.J. (eds.), Information Systems Design Methodologies: A Feature Analysis. Olle, H.G. Sol, C.J. Tully (eds.), Information Systems Design Methodologies: A Feature Analysis, North-Holland, Amsterdam, The Netherlands (1983)
50. Kensing, F.: Towards evaluation of methods for property determination. In: Bemelmans, M.A. (ed.) Beyond Productivity: Information Systems Development for Organizational Effectiveness, North-Holland, Amsterdam, The Netherlands, pp.325–338 (1984)
51. Wijers, G.M., Heijes, H.: Automated support of the modelling process: a view based on experiments with expert information engineers. In: Steinholtz, B., Sølvberg, A., Bergman, L. (eds.) Advanced Information Systems Engineering. CAiSE 1990. Lecture Notes in Computer Science, vol. 436, pp 88–108. Springer, Berlin (1990). https://doi.org/10.1007/BFb0000588
52. Pressman, S.R.: Software Engineering. A practical approach, 7th edn, Mc Graw Hill, Mexico City (2010)
53. IEEE Recommended Practice for Software Requirements Specifications, IEEE Std 830-1998 pp.1–40 (1998)
54. Martin, J.N.: Overview of the EIA 632 standard: processes for engineering a system. In: 17th DASC. AIAA/IEEE/SAE. Digital Avionics Systems Conference. Proceedings (Cat. No. 98CH36267), pp. B32–1 (1998)

Databases and Data Mining

Distribution Analysis of Postal Mail in Argentina Using Process Mining

Victor Martinez[1]($\boxtimes$)(iD), Laura Lanzarini[2](iD), and Franco Ronchetti[2,3](iD)

[1] School of Computer Science, National University of La Plata, La Plata, Argentina
martinezvictor@hotmail.com
[2] Computer Science Research Institute LIDI (III-LIDI) (UNLP-CICPBA),
La Plata, Argentina
[3] Scientific Research Agency of the Province of Buenos Aires (CICPBA),
La Plata, Argentina

Abstract. Process mining combines a number of techniques that allow analyzing business processes solely through event logs. This article is a continuation of the research carried out in [1] to analyze data based on the postal distribution of products in the Argentine Republic between the years 2017 and 2020. The results obtained initially showed that 85% of the shipments made comply with the process correctly. Cases that did not fit within the model were also quickly identified, and recurring problems were found, which facilitates analysis for process improvement. The most common problems were traces that do not follow task order, excess movements or missing movements, and traces that comply with the process but take too long. In this article, a performance analysis was added to discover traces that, despite correctly following to the process, have operational deviations due to an excessive time to complete. These techniques are intended to be added to the process through early alerts that warn about the existence of such situations, which would help improve service quality.

Keywords: Process mining · Data mining · Postal distribution · Postal processes · Business process management

1 Introduction

Currently, there is a wide variety of information systems that support the various areas in each company, both business-related and administrative. These tools include ad hoc developments for each company and management tools such as CRM, ERP, WMS, TMS, BI, etc.

Nowadays, storing data can be done easily and with a low cost, so most of these systems save information in text files or databases for auditing purposes or for error resolution. Among other things, it logs what was done, when it was done, who did it, if there were any errors, etc. Analyzing this data, insights

P. Pesado and G. Gil (Eds.): CACIC 2021, CCIS 1584, pp. 159–169, 2022.
https://doi.org/10.1007/978-3-031-05903-2_11

can be obtained about errors or problems that have occurred, the business process carried out, and how to propose improvements or find solutions to various operational issues.

Process mining provides techniques that allow these logs to be analyzed to obtain insights.

Process mining can answer questions such as: How can process control be improved? What actually happened? Why did it happen? What could happen in the future? How can the process be improved to increase performance? [2].

By working directly with productive data, the real behavior that is carried out for the business and the complete process that is being carried out are obtained, which in some cases may differ from the one originally designed.

The case study in this work is the distribution of products by postal mail. It is a process in which different movements are logged, made up of events such as receiving a product at a branch, its entry into a distribution center, internal movements and the various delivery attempts. All tasks are clearly defined and must be performed in a certain order and within a given time interval. In daily operations, deviations occur due to delays in the execution of tasks or inconsistency (repeated tasks or tasks carried out in an incorrect order).

There are some jobs that link the postal business, process mining and big data, such as [3], where data mining is applied to China mail in a big data environment. Given the complexity of the postal business, clustering techniques were used to group customers based on consumption habits, main interests and behavior, so as to achieve a more effective and accurate marketing strategy. The results were very satisfactory. Another example is [4], where process mining is applied to a logistics and manufacturing chain to look for the similarities and differences between various delivery processes in a changing environment. To do this, different processes are compared using clustering techniques to automate process documentation. Finally, in [5], a methodology that can be used as a guide in process mining projects is shown, and the case study of its application in IBM is discussed as an example.

This work continues the research published in [1] in relation to the application of process mining techniques to postal distribution in Argentina to analyze its operation, identify deviations or issues in distribution, and propose enhancements that will help improve service quality. To add to this previous work, a performance analysis has been added that allows identifying a new type of operational deviations. The data analyzed correspond to postal mail distribution in Argentina between 2017 and 2020. It should be noted that, at the date of publication of this article, the authors are not aware of the existence of any other works with these characteristics for the postal business.

2   Process Mining

Process Mining operates on the log files of the information system to be analyzed. It begins with the extraction of the information of all the events of a certain activity and then, through an automatic analysis, the relevant business process is identified, as well as its activities and the sequence that has to be followed. This allows analyzing and answering questions such as: What happened? Why did it happen? What could happen in the future? How can control be improved? Can performance be improved? [2]. With the information obtained after model interpretation, different types of analysis can be carried out, such as who carries out the activity, whether the activity is complete and if it is carried out within a reasonable time interval. Process mining can be classified into three types [6]:

- *Discovery*: These are techniques capable of modeling the business process that is being carried out solely from the sequence of corresponding logs, which can be extracted from files, databases or some other media. Possibly the most popular algorithm to perform this task is the Alpha algorithm [7], which generates a Petri net with the discovered process.
  The quality of the generated model is directly proportional to the quality of the input data. This is a feature common to any inductive knowledge extraction technique. It is essential that the event sequences surveyed fully represent the process to be modeled; otherwise, there will be aspects that will remain undiscovered.
- *Compliance Verification*: It allows contrasting a real sequence of events against a process model (it can be the one previously discovered or a different one) to determine the compliance level and which are the occurrences that deviate from the process.
  Applications that graphically represent the results and perform animations can be used to liven up the presentation of the results. The resulting analysis can be very accurate, since it is based on real data rather than being a simulation.
- *Improvement*: Its goal is improving the existing process based on the analysis carried out. It differs from Compliance Verification in that it focuses on the process and not on trace execution.

There are four other aspects to take into account to obtain quality results in process mining. These are accuracy, fit, generalization, and simplicity [2]. A balanced relationship between these four forces has to be maintained so that the discovered model is representative of the process and easy to understand.

Last but not least, it should be noted that input data usually carry a large amount of noise, which can be due to data duplication or incomplete traces that can distort analysis results [8]. As in other mining techniques, pre-processing is usually carried out to improve input data quality by removing incomplete or corrupt data that could lead to errors in modeling.

3 Process Mining Applied to Postal Distribution

The discovery of the process will be carried out using data from the postal distribution in Argentina between 2017 and 2020.

Postal mail can be used to distribute various products, such as letters, telegrams, parcels or, less traditionally, e-commerce products.

In all cases, during the process, at least the following steps take place: regis-tration and reception of the product to be distributed, internal routing through one or more sites or distribution centers, one or more delivery attempts, effective delivery or return to sender if delivery was not successful. Each step carried out is always logged, including information about the person responsible for the action, what they did, and when they did it. This record is associated with a unique shipping identifier that allows these events to be reported to the customer.

3.1 Data Extraction

To discover the model and then analyze the events, data will be extracted and pre-processed. For this particular case study, product shipments that required two delivery attempts were used. The procedure in this case is as follows: the product goes to distribution, if it cannot be delivered for any reason that is not final, a new delivery is attempted the next day. If it cannot be delivered once again, a period of time is given to the recipient to come and pick it up (a visit notice is left at each visit). Once the established period of time has passed, if the recipient has not collected the shipment, it is returned to the sender.

Each shipment is considered as a trace and each movement as an event. The trace is considered to be completed when the shipment has an entry and an end on record, with successful delivery or not.

As a result of data collection, a sample containing around 33,000 traces with more than 77,000 events was generated (see Fig. 1). Each trace has at least two associated events: a trace identifier and an identifier for each event. In particular, the necessary fields for analysis were recorded for all the events in the case study: trace identifier, event identifier within the trace, description, and event date.

With this information, each trace can be rebuilt and the corresponding model put together.

	Trace ID	Event ID	Event description	Event date
	123 trazaID	123 EveID	ABC eveDescrip	eveFecha
31	481,053	0	INGRESADO	2017-08-16 10:45:22
32	481,053	2	1 INTENTO DE ENTREGA	2017-08-18 11:15:00
33	481,053	9	DEVOLUCION	2017-08-18 13:00:00
34	481,054	0	INGRESADO	2017-08-16 10:45:28
35	481,054	2	1 INTENTO DE ENTREGA	2017-08-22 12:05:00
36	481,054	9	DEVOLUCION	2017-08-22 17:25:00
37	481,055	0	INGRESADO	2017-08-16 10:45:27
38	481,055	1	ENTREGADO	2017-08-22 15:13:00

Fig. 1. Example of events extracted for analysis

Data are then converted to XES format [9]. XES proposes a tag-based lan-guage that provides a unified and extensible methodology to log behaviors in information systems [9]. Data structured in this way can be processed with a wide variety of tools in an efficient way.

To increase data consistency and facilitate the analysis, all incomplete traces—either because they did not have an initial status (entry) or a final status (delivered, returned, no address exists, died, or moved)—were removed. This may have been due to an error when loading the data or because the traces have not yet completed the process.

To remove incomplete traces, a simple heuristic rule filter was used, where the initial state and the valid final states were specified. The filter discards any trace that does not meet these requirements. After applying the filter, a sample of approximately 16,000 traces with 43,000 events was obtained.

3.2   Process Discovery

The Alpha algorithm was used for process discovery. The Alpha algorithm was proposed by van der Aalst, Weijters and Maruster [7], and is widely used in process mining. The algorithm rebuilds causality from a sequence of events and returns a Petri net where the business process used is reflected. Each transition in the network represents a task.

For this particular case study, a small representative sample is extracted that will be used to discover the process; then, the discovered process is contrasted with the rest of the traces to observe compliance.

Figure 2 shows the discovered Petri net that represents the process that the traces must follow.

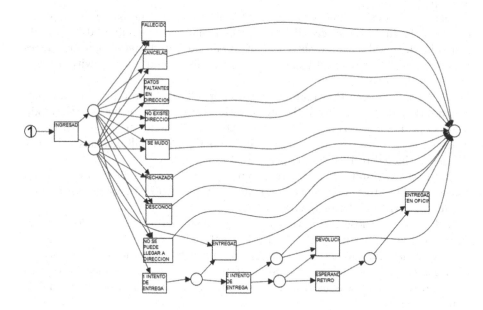

Fig. 2. Process discovered that all traces must fulfill

The discovered process is made up as follows: there is always an entry event that all the traces have to complete, which is followed by the distribution step. If delivery is possible, it is logged and the history of the trace ends. If it is not possible due to a final event (deceased person, the address does not exist, there is not enough data to find the address, etc.), a non-delivery event is logged with the relevant reason and the process ends. If the reason for non-delivery is not a final event (for example, there was no one to receive the parcel), a first delivery attempt is logged and another attempt is made the following day. If on the following day it cannot be delivered once again, a period of time is given to the recipient to go to the office and pick it up (a visit notice is left at each visit). Finally, the delivery event is logged as delivery in the office or the shipment is returned to the sender. Figure 2 illustrates the scenarios of the discovered process.

3.3 Model Verification

To check compliance, all sample traces are taken (a total of 16,811 traces with 43,888 events) and they are compared to the discovered model.

The aim is to determine how much the traces fit the model and analyze those that deviate the most, taking into account which steps are fulfilled, which are not, and if there are steps that are not reflected in the discovered process. The result is shown in Fig. 3, with the most common events highlighted in a dark color and the most frequent paths represented with a thicker line. It can be seen that most of the traces end with the delivery, either on the first or the second attempts.

Sample statistics indicate that in 80% of the cases, the parcel is delivered either in the first or the second attempt, and that the remaining 20% is evenly distributed. Based on these observations, it can be stated that 85% of the traces (either delivered or not) fully comply with the discovered process, with a mean of 2.6 events in each trace.

It can also be seen that many traces do not follow the process exactly, either because they do not complete the steps in the correct order or because they skip some step. The more they deviate from the model, the lower the match to the process; therefore, those traces with a fit of less than 50% will be analyzed, since this is considered to be a large operational deviation.

With these results, two different types of analyses can be carried out: 1) traces that have too many movements, and 2) traces whose tasks are not carried out in the correct order or have missing events.

Additionally, those traces that correctly follow the process but do so in an excessively long time interval will also be analyzed.

Given that the number of movements per trace follows a normal distribution with a mean of 2.6 and a deviation of 0.95, the value of the mean plus two standard deviations was used as a representative value of an excessive number of movements.

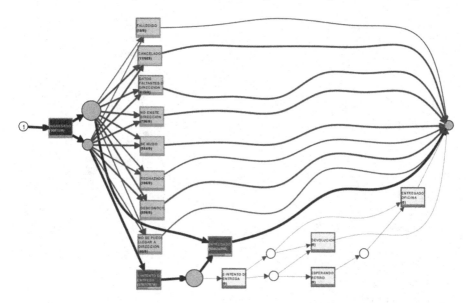

Fig. 3. Trace paths: most common events highlighted in a dark color and most frequent paths represented with a thicker line.

Based on this, in the first case, traces with more than 6 movements were filtered and exported for further analysis. A total of 108 cases with these characteristics were found.

In the second case, the traces that did not exactly match the process discovered were analyzed. For this, an adjustment value of 0.5 was used, considering that traces have at least 50% of their movements outside of the model, representing a large deviation from the standard procedure. As previously mentioned, this can happen because there are too many events in the traces (they do not exist in the model) or they do not correctly follow the sequence.

In this case, it was observed that the same event is repeated on successive days or there are inconsistent logs. Figure 4 shows these situations. On the left side, a case with a first visit after the second is shown. This is impossible, meaning it is probably due to a logging error (for example, the wrong date was used in one of the movements). On the right side, in that same figure, an inconsistency is shown where a delivery movement is logged after the product was returned.

Additionally, the Inductive Visual Miner tool [10] was used to generate a visual animation and represent these cases in a more friendly way. For this, the set of traces with a match rate of less than 50% with the process was used, obtaining the visualization of the traces that deviate the most. Figure 5 illustrates this animation. In this figure, each trace is represented with a circle that goes through the different stages of the process. This visual representation shows that some traces take much longer than others. It also shows that, sometimes, there are incorrect behaviors. For example, in Fig. 5, traces go back, i.e., after a task is completed, a previous task is carried out; circles are used to highlight

Fig. 4. Repeated events and inconsistencies

that 25 traces go from the second delivery attempt to the first one, and 20 repeat the same step.

The export of traces that do not follow the model shows the reasons behind the deviations, which allows carrying out a more detailed analysis to identify potential improvements to the process.

3.4   Performance Analysis

Finally, a performance analysis is carried out on the traces in the model to identify those that, although they may have completed the process correctly, did so in a much longer time than the mean resolution time. To do this, the Petri net data sample was replicated using a tool oriented to performance analysis that focuses on resolution times and not so much on seeing whether the traces follow the process or not.

As a result, a visualization is obtained that represents in darker tones the tasks that take more time to complete.

A mean resolution time for each trace of 3.75 days is observed, with a deviation of 2.81 days. Using the same criteria as with the number of movements per trace, the analysis now focuses on those cases that have a completion time greater than the mean plus two deviations (9.37 days), considering that a resolution time greater than this value is an operational diversion.

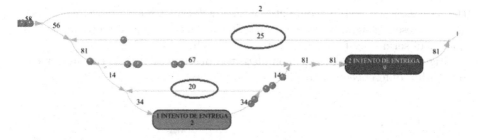

Fig. 5. Traces that do not follow the model; those that go back to a previous state instead of moving forward to the next are circled (see arrow direction)

The traces that meet this criterion are exported, obtaining a total of 687 cases (Fig. 6).

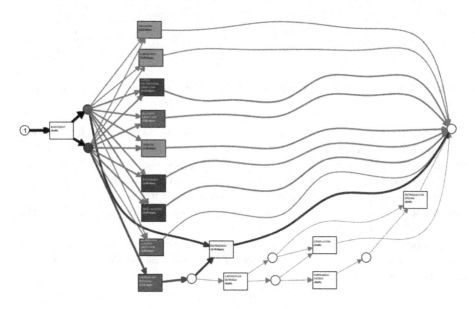

Fig. 6. Performance analysis of sample traces

Among these, there are traces that perfectly follow the process, but with extremely long times. In some cases, traces have more than three movements and an interval of up to 3 days between them, which explains such a long overall time. Even though 3 days between movements is longer than the mean, this could be due to situations that may occur on a day-to-day basis, such as lack of staff due to illness or vacation. On the other hand, most of the cases in this sample have only two movements, where entry was logged and then a period of up to 14 days passed before an exit to distribution event was logged. After this, the process is successfully completed with a delivery or return movement. These cases represent a significant operational deviation, since they directly impact the quality of service by failing to meet product delivery standards.

Figure 7 shows traces that complete the process with only two movements but with a difference of more than 10 days between them.

3765390.0 2 events	3769366.0 2 events	3763891.0 2 events	3763937.0 2 events
INGRESADO #1 07.01.2020 10:58:28.000	INGRESADO #1 10.01.2020 16:30:40.000	INGRESADO #1 03.01.2020 20:23:55.000	INGRESADO #1 03.01.2020 20:40:43.000
ENTREGADO #2 22.01.2020 09:55:00.000	ENTREGADO #2 28.01.2020 10:32:00.000	ENTREGADO #2 17.01.2020 13:00:00.000	DESCONOCIDO #2 16.01.2020 16:37:00.000
3764216.0 2 events	3769606.0 2 events	3767176.0 2 events	3775013.0 2 events
INGRESADO #1 03.01.2020 20:45:15.000	INGRESADO #1 06.01.2020 20:10:58.000	INGRESADO #1 07.01.2020 15:56:15.000	INGRESADO #1 15.01.2020 19:53:23.000
ENTREGADO #2 23.01.2020 13:20:00.000	ENTREGADO #2 21.01.2020 10:31:00.000	ENTREGADO #2 23.01.2020 12:01:00.000	SE MUDO #2 27.01.2020 12:30:00.000

Fig. 7. Traces that follow with the process but take a very long time to complete

The performance analysis helped uncover new operational deviations that had not appeared in the previous analysis. These are traces that perfectly follow the model but whose completion times are well above the mean, and sometimes can be as long as 24 days before delivery. Based on this study, it was decided to take a sample of the 10 longest cases and ask the heads of the corresponding sector/branch about them, in an attempt to gather information about the reasons behind these delays.

The performance analysis helped identify a need to generate an alarm that brings attention to these cases after a certain time has passed, which would help improve quality of service.

4   Conclusion and Future Lines of Research

In this article, different process mining techniques have been used to analyze real data based on the postal distribution of products in Argentina between 2017 and 2020. At the date of generation of this document, no similar investigations have been found that analyze this same case study.

After generating the model from a representative sample of traces, the process that actually takes place was discovered. Subsequently, this model was contrasted with all traces. Through a compliance check, unusual situations that resulted in operational errors were identified. Cases that did not comply with the model were detected, as well as cases that presented task redundancy.

The performance analysis allowed discovering traces that correctly follow the process but have operational deviations due to an excessive time to complete. These cases have a great impact on service quality.

As a future line of work, process mining techniques will continue to be used in an attempt to find a way to insert early warnings into the system aimed at avoiding deviations and operational bottlenecks.

References

1. Martinez, V., Lanzarini, L., Ronchetti, F.: Process mining applied to postal distribution. In: CACIC 2021 (2021). http://sedici.unlp.edu.ar/handle/10915/130342
2. van der Aalst, W.: Process Mining: Data Science in Action, 1st edn. Springer, Heidelberg (2016). isbn: 978-3-662-49850-7.break https://doi.org/10.1007/978-3-662-49851-4
3. Hu, X., Jin, Y., Wang, F.: Research of postal data mining system based on big data. In: 3rd International Conference on Mechatronics, Robotics and Automation (2015). https://www.researchgate.net/publication/300483008_Research_of_Postal_Data_mining_system_based_on_big_data
4. Tseng, M.M., Tsai, H.-Y., Wang, Y.: Context aware process mining in logistics. In: The 50th CIRP Conference on Manufacturing Systems (2017). https://www.sciencedirect.com/science/article/pii/S2212827117303311
5. van Eck, M.L., Lu, X., Leemans, S.J.J., van der Aalst, W.M.P.: PM2: a process mining project methodology. In: Eindhoven University of Technology, The Netherlands (2017). http://www.processmining.org/_media/blogs/pub2015/pm2_processminingprojectmethodology.pdf
6. van der Aalst, W.: The process mining manifesto by the IEEE task force. In: IEEE Task Force (2012). https://www.tf-pm.org/resources/manifesto
7. van der Aalst, W., Weijters, T., Maruster, L.: Workflow mining: discovering process models from event logs. IEEE (2004). https://ieeexplore.ieee.org/document/1316839
8. Gunther, C.W.: Process mining in flexible environments. In: Technische Universiteit Eindhoven (2004). https://research.tue.nl/en/publications/process-mining-in-flexible-environments
9. IEEE Std 1849-2016: IEEE Standard for eXtensible Event Stream (XES) for Achieving Interoperability in Event Logs and Event Streams. IEEE (2016). https://doi.org/10.1109/IEEESTD.2016.7740858
10. Leemans, S.J.J.: Inductive visual miner (2017). http://leemans.ch/leemansCH/publications/ivm.pdf

Anorexia Detection: A Comprehensive Review of Different Methods

María Paula Villegas[1,2]($\boxtimes$) (iD), Leticia Cecilia Cagnina[1,2](iD),
and Marcelo Luis Errecalde[1](iD)

[1] Laboratorio de Investigación y Desarrollo en Inteligencia Computacional,
Universidad Nacional de San Luis (UNSL),
Ejército de los Andes 950, 5700 San Luis, Argentina
villegasmariapaula74@gmail.com
[2] Consejo Nacional de Investigaciones Científicas y Técnicas (CONICET),
San Luis, Argentina

Abstract. The need for identity validation and self-image approval by
the society places adolescents and young adults in a situation of vul-
nerability. Social networks can make this validation positive, but they
could also be a risk factor that triggers various Eating Disorders (ED),
particularly, Anorexia Nervosa. Many technologies have already started
trying to identify when these risks exist. Thus, our main objective in this
work is to analyze the performance of various methods that allow the
early detection of anorexia nervosa. In principle we analyze the performance of representations such as k-TVT, Word2Vec, GloVe and BERT's
embeddings, classifying with standard algorithms such as SVM, Naïve
Bayes, Random Forest and Logistic Regression. Then, we carry out an
analysis of the performance of the classifying models comparing those
classical models with methods based on deep learning, such as CNN,
LSTM and BERT. As a result, k-TVT, CNN and BERT's embeddings
performed best.

Keywords: Anorexia early detection · Learned representations ·
Classical text representations · Temporal variation of terms

1 Introduction

The World Health Organization (WHO)[1] estimates that worldwide 70 mil-
lions people have an eating disorder such as anorexia nervosa, bulimia nervosa,
orthorexia, binge-eating disorder or related conditions that put their mental and
physical health at risk. These disorders commonly emerge during adolescence
and young adulthood, and involve abnormal eating behavior, accompanied in
most cases by concerns about body weight and shape.

Experts agree that the causes of eating disorders can be sociocultural, psy-
chological, hereditary, among others. However, these disorders are closely related

---

[1] https://www.who.int/news-room/fact-sheets/detail/adolescent-mental-health.

P. Pesado and G. Gil (Eds.): CACIC 2021, CCIS 1584, pp. 170–182, 2022.
https://doi.org/10.1007/978-3-031-05903-2_12

to self-image, and it is the exaltation of a perfect image, one of the most negative trends in social networks, which makes the problem worse [15]. Adolescents and young people are the ones who use regularly social networks and, in the search for their identity, they are particularly vulnerable to their content.

Anorexia nervosa can lead to premature death, often due to medical complications or suicide, and has higher mortality than any other mental disorder. For this reason, it emphasizes the importance of early detection of this type of problem.

We previously studied the suitability of state-of-the-art representations including: k-TVT, GloVe, Word2Vec and, BERT's embeddings for early detection of anorexia risk [19]. These representations were chosen because they are considered standard on many problems, including the one studied here [1,3,14,17,20]. However, we performed experiments using classical classification models based on: Support Vector Machines (SVM), Naïve Bayes (NB), Random Forest (RF), and Logistic Regression (LR).

In this extended version of [19] we will try to answer the following research questions:

– **Research Question 1**: How different text representations perform in the detection of anorexia?
– **Research Question 2**: Do deep learning methods work better for the studied task?
– **Research Question 3**: In detecting signs of anorexia, which state-of-the-art model works best?

Therefore, in Sect. 2 we briefly define the text representations and classification methods that have been considered in this article. Then, in Sect. 3, the dataset used in the experiments is shortly described along with the description of the early detection of anorexia signs task. In Sect. 4 the experimental study is carried out. Sections 4.1 and 4.2 discuss methods using standard classifiers. Sections 4.3 and 4.4 analize the results obtained using deep learning. Finally, Sect. 5 summarizes the main conclusions and future work.

2   Analysed Methods

In [19], classical representations such as Bag of words and n-grams of characters, k-TVT and the learned representations Word2Vec, GloVe and, BERT's embeddings were selected. We provide a brief description of each below. Finally, a short introduction about the deep learning classification methods (BERT, CNN and LSTM) used in this task is provided.

Bag of Terms (BoT) is a widely used representation that represents each document as a bag, that is, an unordered set of terms. We will refer to 'term' as a word, as a sequence of two or more words, or as a sequence of two or more characters. The first case, **Bag of Words (BoW)** [8], is simple to implement and fast to obtain. However, it has the disadvantage of losing a lot of semantic and conceptual information by ignoring the word order in the document. In the second case, a term is considered as a sequence of words or characters, obtaining another representation called **n-grams of words** or **n-grams of characters** respectively [5]. In the English language particularly, the 3 g of characters are the ones that have shown the best performance [2] because, among other things, they are useful for informal text collections where misspelled words or words with repetitions of characters tend to appear. Formally, in BoT, a document is represented by a weight vector of each term, according to the chosen weight scheme (tf, tf-idf, boolean, among others). The vectors in this representation are usually huge and can be very sparse.

On the other hand, k-**TVT** [4] is a representation that, in addition to words, focuses on the context in which they occur. This type of approach is based on the principle that words that occur in similar contexts tend to have similar meanings [8]. Each context can be modelled through semantic elements called concepts and thus, represent words and documents with a combination of concepts. The set of concepts associated with each word can be viewed as a bag of concepts and then the document will be represented by the concepts associated with the words in that document.

In particular, k-TVT first represents the terms based on the different contexts in which they can occur, and then generates the document vectors from those representations. To determine the concepts, k-TVT does it in a simple way: using the class labels of the classification task. In this representation, two words are related if their relative frequency distributions in the documents of the different classes are similar. That is, the more frequent a word is in documents that belong to a class, the greater its membership in that class. In addition to the good performance shown by the models that use k-TVT, other advantages are the low dimensionality of the vectors that represent each document and the balance that it performs between the minority (or positive) class with respect to the majority (or negative) class.

Finally, the learned representations emerge, where the idea is to extend machine learning, usually used in a later step to generate the classification model, to the document representation. For the learning of representations, word embeddings are used, which are basically distributed representations of words based on dense vectors of fixed length, which are obtained from statistics of the co-occurrence of words according to a distributional hypothesis.

In the specific literature, it is possible to distinguish between representations derived from counting-based approaches such as **GloVe** (*Global Vectors*) [13], and those arising from predictive neural learning methods such as **Word2Vec** [12] and **BERT** (*Bidirectional Encoder Representations from Transformers*) [6], among others. In predictive approaches, neural networks with many units are used and they are fed with extensive collections of texts in an unsupervised way, which enables representations to learn general concepts of languages. Thus, word embeddings capture very interesting syntactic and semantic relationships of words, such as relational meanings.

Word2Vec is a method used to obtain distributed representations of words (word embeddings). Basically, the authors in [12] proposed to learn a classifier in order to obtain word representations as a collateral effect. Embeddings are extracted from the pre-output layer of a neural network classifier. This approach considers the context in which the word is found using the size of the context window as a parameter. For small windows, the words most similar to the target word will be semantically similar and will have the same grammatical category. On the other hand, in larger windows the words that are most similar to the target word will be semantically related without necessarily becoming similar.

The local context window parameter in combination with a global matrix factorization determines the representation of **GloVe**. The authors in [13] argue that the quotient of the probabilities of coexistence of two words is what contains the relevant information. First, the co-occurrence matrix between all words in the vocabulary in the text collection is calculated. Then, based on that matrix, the algorithm learns a vector for the representation of each word and another vector in which the context of that word is modelled. Finally, both vectors can be averaged to obtain the vector representation of each word and then, as in Word2Vec, they will be used to represent a document.

In recent years, the state of the art in the field of natural language processing has focused on predictive approaches based on transformers [18]. Transformers are a type of deep neural network architecture that includes an attention mechanism. These mechanisms encode each word of a sentence as a function of the rest of the sequence, thus allowing context to be introduced into the representation (contextual embeddings).

BERT is a model created by Google in 2018 [6] with the intention of solving certain Natural Language Processing (NLP) tasks. The authors basically proposed, instead of generating a model that solves each task individually, to train a base model that learns to interpret the language in general and then add some additional layers that allow the model to specialize in a particular task.

BERT is a model created by Google in 2018 [6] with the intention of solving certain Natural Language Processing (NLP) tasks. The authors basically proposed, instead of generating a model that solves each task individually, to train a base model that learns to interpret the language in general and then add some additional layers that allow the model to specialize in a particular task. This model pre-trains deep bidirectional representations from unlabeled text and can be fine-tuned with just adding an output layer to the neural network of the model. BERT can process a document as a sequence of sentences of tokens; analyzing left and right contexts of each token it produces a vector representation for each word as the output, considering a pre-trained model. BERT is pre-trained on a large corpus of unlabeled text: 2,500 million words extracted from Wikipedia and 800 million from Book Corpus. Although we can perform the fine-tuning of the model on a specific task and task-specific data, in this work we use the BERT's embeddings (BEMB) as the representation in Sects. 4.1 and 4.2, while we use the complete classification model in Sects. 4.3 and 4.4.

A **convolutional neural network (CNN)** is a class of artificial neural network, commonly used to analyze visual images [10]. These networks do a task similar to that of the human brain when processing an image. At the input you have an image in digital format and the first layers of that Convolutional Network extract basic patterns, such as lines and edges, and little by little, as we go deeper into the network, these basic elements are combined in increasingly complex shapes and figures until finally the model is capable of detecting the image. The Convolutional Network takes a digital input data and, progressively, through the training of multiple convolutional layers (made up of filters that perform different operations on the image), increasingly complex characteristics of the image are progressively extracted to achieve the classification learning. Subsequently, the output of these filters is taken to deeper layers called **max-pooling** that allow reducing the amount of information, and thus extract the most representative data. Then, these procedures (filtering and max pooling) are repeated and each time the size of the resulting images becomes smaller. In the end, few layers allows taking the features extracted by the convolutional layers, representing them as a data vector and making the final classification of the image, using a **softmax** or **sigmoidal function**.

On the other hand, **Recurrent Neural Networks** [16] are capable of processing different types of sequences (such as videos, conversations, texts). These types of networks are characterized by having certain memory limitations. However, within Recurrent Neural Networks, there is an architecture widely used today: the **LSTM (Long Short Term Memory) Networks** [9], which solve the problem of memory limitations, because they can add or delete information, keeping what they consider relevant for the processing of the sequence.

In Sect. 4, we will detail the implementation of each of these methods.

3   Data Set and Pilot Task

As in the previous article [19], here we consider early prediction only of the anorexia disorder [7]. To do this, we have worked with a collection of posts or comments from Reddit social networks that was provided by the eRisk 2018 laboratory. Let us take into account that there are two categories of users: those diagnosed with the disorder, in this case anorexia (positive class) and the group that does not suffer from it or the control group (negative class). What is written by each user is divided into 10 fragments. This sequence of writings is ordered chronologically. That is, the first snippet contains the oldest 10% of comments, the second snippet contains the second oldest 10%, and so on. The following table (see Table 1) summarizes, for both sets, the number of users in each class.

As for the Early Detection of Anorexia Signs task, this consists of determining if a user presents any signs compatible with the aforementioned disorder, indicating whether or not the person is at risk of suffering from it. The challenge is to sequentially process the texts of each individual and detect the first traces of anorexia as soon as possible.

Table 1. Data set for the anorexia detection task.

Training		Test	
Anorexia	Non-anorexia	Anorexia	Non-anorexia
20	132	41	279

4 Experimental Study

In this study we use the corpus described above. First, in Subsect. 4.1, the results were obtained with the classical classification methods: Support Vector Machine (SVM), Naïve Bayes (NB), Random Forest (RF) and Logistic Regression (LR) and the selected representation: Bag of Words (BoW), character trigrams (C3G,) k-TVT, Word2vec (w2v), GloVe and BERT's embeddings (BEMB). Then, in Subsect. 4.2 we show a comparison between the results of the previous section with those obtained in eRisk 2018 competition.

In Subsect. 4.3 we show the results obtained with the deep learning models: CNN, LSTM and BERT. Finally, in Subsect. 4.4 we briefly analyze the performance of all the experimented methods in this task.

The executions were carried out using programming language Python 3, on the on-line Google Colaboratory platform. The implementations of the classical classification algorithms correspond to those provided in the *Python scikit-learn* library with the default parameters.

The performance of the classifiers was evaluated using the F1, precision, recall and early risk detection error ($ERDE_\theta$) [4] measures. $ERDE_\theta$ simultaneously evaluates the precision of the classifiers and the delay in making a prediction. Delay refers to the amount of posts to be considered before the classifier takes a decision. The parameter θ is a time limit for decision-making.

Regarding the word embedding models Word2Vec and GloVe, we use the pre-trained vectors[2], given the computational effort required to obtain new ones. Then, the document vectors were obtained by averaging the embeddings of the words that appear in each one (we use vectors of 300 dimensions).

In the case of BEMB, the texts of the users were divided into sentences and the bert-as-service[3] library was used to extract the corresponding vectors from the model. Then, the document text representation vector was generated by averaging the embeddings of the sentences which have 768 characteristics. We use BERT pre-trained vectors corresponding to the *Base* version available for download in the official Google repository[4].

---

[2] Pre-trained vectors were obtained from https://nlp.stanford.edu/projects/glove/ and https://s3.amazonaws.com/dl4j-distribution/GoogleNews-vectors-negative300.bin.gz.

[3] https://github.com/hanxiao/bert-as-service.

[4] https://github.com/google-research/bert.

4.1 Performance of the Different Text Representations with Standard Classifiers

Table 2 shows the comparison of the results obtained with each model. Regarding the baseline (shown in Table 2 in italic), we have taken the classical representations such as bag of words (BoW) and trigrams of characters (C3G) and we have selected the best of those as reference value. For both, different weighting scheme were used: Boolean, Term-Frequency (TF) and Term Frequency - Inverse Document Frequency (TF-IDF), where the best results were achieved with the latter.

As we said before, k-TVT defines concepts that capture the sequential aspects of early risk detection problems and the vocabulary variations observed in the different stages of writings. Therefore, a different number k of chunks that will enrich the minority (positive) class could have an impact on the $ERDE_\theta$ measure. As for detecting depression, the value of k was varied in the range $[0, 5]$ (integer) and we show the best value obtained with that specific k for each metric.

In each chunk, classifiers usually produce their predictions with some confidence, generally the estimated probability of the predicted class. Therefore, we can select different thresholds τ considering that an instance is assigned to the target class when its associated probability p is greater (or equal) than a certain threshold τ $(p \geq \tau)$. Four different scenarios were considered for the assigned probabilities for each classifier: $p \geq 0.9, p \geq 0.8, p \geq 0.7$ and $p \geq 0.6$. Remember that once a classifier determines that an instance is positive in a specific chunk, that decision remains unchanged until chunk 10.

In the tables, we can see the notation X-TVT, where X references to number of chunks (k) used for the representation k-TVT. For the learned text representations, we use Word2Vec (w2v), GloVe and BERT acronyms. In Table 2 we can observe the representations in the first column, then a column that refers to the classifier used (SVM, NB, RF and LR), and the probability of the predicted class in the third column. The following columns correspond to the metrics. Due to space limitations, we selected for each metric, those configurations that obtained the best values with each text representation. Thus, for example, the first five rows correspond to the best $ERDE_5$ for each configuration. In addition, we wanted to emphasize the best value among the best ones, for which we highlighted in bold the model (representation with classifier-probability configuration) that performs best. In this case, $ERDE_5$ is BoW with LR and probability 0.6.

We can observe that k-TVT obtained good results for all $ERDE_\theta$ (quite similar in $ERDE_5$ to the best that is the baseline). We found that an adequate configuration of the k parameter is: $k = 1$ when the level of urgency is high ($ERDE_5$ and $ERDE_{10}$) and $k = 5$ (more information is taken into account in the representation) when the level of urgency is lower ($ERDE_{25}$, $ERDE_{50}$ and $ERDE_{75}$). Its combination with the Random Forest classifier and low probabilities are suitable configurations for obtaining good $ERDE_\theta$ metrics.

Table 2. Comparison between all models for early detection of anorexia (standard classifiers)

		Classifier	p	$ERDE_\theta$					F1	Pre	Re
				$\theta = 5$	$\theta = 10$	$\theta = 25$	$\theta = 50$	$\theta = 75$			
Best $ERDE_5$	BoW	LR	0.6	**11.52**	11.14	10.23	8.68	8.68	0.62	0.78	0.51
	1-TVT	RF	0.6	11.55	**10.58**	9.18	7.47	7.47	0.67	0.77	0.59
	w2v	SVM	0.7	11.77	11.37	10.28	8.72	8.54	0.50	0.70	0.39
	GloVe	SVM	0.6	11.67	11.22	9.68	7.55	7.51	0.62	0.73	0.54
	BEMB	RF	0.6	11.68	11.08	9.61	7.67	7.37	0.66	0.68	0.63
Best $ERDE_{10}$	BoW	LR	0.6	**11.52**	11.14	10.23	8.68	8.68	0.62	0.78	0.51
	1-TVT	**RF**	**0.6**	11.55	**10.58**	9.18	7.47	7.47	0.67	0.77	0.59
	w2v	LR	0.6	11.83	11.15	9.94	8.52	8.37	0.53	0.59	0.49
	GloVe	SVM	0.6	11.67	11.22	9.68	7.55	7.51	0.62	0.73	0.54
	BEMB	RF	0.6	11.68	11.08	9.61	7.67	7.37	0.66	0.68	0.63
Best $ERDE_{25}$	C3G	SVM	0.6	11.64	11.26	10.03	8.49	8.19	0.59	0.70	0.51
	5-TVT	**RF**	**0.7**	11.96	10.82	**9.07**	7.36	7.13	0.68	0.69	0.66
	w2v	LR	0.6	11.83	11.15	9.94	8.52	8.37	0.53	0.59	0.49
	GloVe	SVM	0.6	11.67	11.22	9.68	7.55	7.51	0.62	0.73	0.54
	BEMB	RF	0.6	11.68	11.08	9.61	7.67	7.37	0.66	0.68	0.63
Best $ERDE_{50}$	BoW	RF	0.6	12.79	12.08	10.03	7.53	7.27	0.74	0.93	0.61
	5-TVT	**RF**	**0.7**	11.96	10.82	9.07	**7.36**	7.13	0.68	0.69	0.66
	w2v	RF	0.6	12.01	11.34	10.11	8.08	7.94	0.58	0.66	0.51
	GloVe	SVM	0.6	11.67	11.22	9.68	7.55	7.51	0.62	0.73	0.54
	BEMB	RF	0.6	11.68	11.08	9.61	7.67	7.37	0.66	0.68	0.63
Best $ERDE_{75}$	BoW	RF	0.6	12.79	12.08	10.03	7.53	7.27	0.74	0.93	0.61
	5-TVT	**RF**	**0.6**	12.16	11.02	9.27	7.56	**7.09**	0.64	0.61	0.66
	w2v	RF	0.6	12.01	11.34	10.11	8.08	7.94	0.58	0.66	0.51
	GloVe	SVM	0.6	11.67	11.22	9.68	7.55	7.51	0.62	0.73	0.54
	BEMB	RF	0.6	11.68	11.08	9.61	7.67	7.37	0.66	0.68	0.63
Best F1	BoW	RF	0.6	12.79	12.08	10.03	7.53	7.27	0.74	0.93	0.61
	5-TVT	LR	0.6	11.80	10.67	9.22	7.51	7.28	0.68	0.76	0.61
	w2v	RF	0.6	12.01	11.34	10.11	8.08	7.94	0.58	0.66	0.51
	GloVe	SVM	0.6	11.67	11.22	9.68	7.55	7.51	0.62	0.73	0.54
	BEMB	**NB**	**0.6**	13.08	12.59	9.93	7.75	7.51	**0.76**	0.71	**0.83**
Best precision	BoW	RF	0.9	12.81	12.81	12.58	12.50	12.50	0.26	**1.00**	0.15
	2-TVT	RF	0.9	12.40	12.22	11.02	10.71	10.71	0.41	0.85	0.27
	w2v	**RF**	**0.9**	12.81	12.81	12.81	12.81	12.81	0.05	**1.00**	0.02
	GloVe	RF	0.8	12.54	12.20	11.92	11.20	10.67	0.42	0.92	0.27
	BEMB	**RF**	**0.9**	12.81	12.81	12.81	12.50	12.50	0.09	**1.00**	0.05
Best recall	BoW	RF	0.6	12.79	12.08	10.03	7.53	7.27	0.74	0.93	0.61
	5-TVT	RF	0.7	11.96	10.82	9.07	7.36	7.13	0.68	0.69	0.66
	w2v	RF	0.6	12.01	11.34	10.11	8.08	7.94	0.58	0.66	0.51
	GloVe	SVM	0.6	11.67	11.22	9.68	7.55	7.51	0.62	0.73	0.54
	BEMB	**NB**	**0.6**	13.08	12.59	9.93	7.75	7.51	**0.76**	0.71	**0.83**

When we analyse the learned representations, for F1, precision and recall, BEMB obtained the best values. For precision, Bag of Words and Word2Vec also obtained the best value.

As we concluded in previous works, k-TVT is a good alternative to represent documents when urgency is the main factor to be considered in the classification task.

Finally, we can see that learned representations like BEMB, for example, seem to be adequate when precision and recall metrics are considered more important. However, although the results of these representations for the $ERDE_\theta$ measures do not reach the best results, they are close.

Then, if we consider the execution time[5] necessary to obtain each text representation, BEMB takes more time (90 min approximately) than GloVe and Word2vec (25 min approximately); being k-TVT the fastest computed (2 min on average). Regarding the size of the representations (dimension), k-TVT has the lowest dimension. For example, 2-TVT uses vectors of size 4 ($k + 2$) while the others consider large vectors: GloVe and Word2Vec with dimension 300 and BEMB with 768. Therefore, the larger the size, the more computing power is required and, consequently, the response delay is greater.

4.2 Performance Comparison Between Models Using Standard Classifiers with Previous Results

To complement our analysis, we compared the best results in Table 2 with the results obtained in the eRisk2018 laboratory. In the proposed task, only F1, Precision, Recall, $ERDE_5$ and $ERDE_{50}$ were considered; we only show the best values for those metrics.

In the first four rows of Table 3 we can see the best values obtained in the experiments that we are carrying out in this study (BoW best $ERDE_5$, 5-TVT best $ERDE_{50}$, BEMB with NB best F1 and recall, BEMB with RF best precision). While in the last four rows we can see the best values (final results) published by the organizers for each metric reported.

It should be noted that the acronym **UNSL** refers to our participation in the competition, where **UNSLB** achieved the lowest (best) value for the measure $ERDE_5$. While the lowest value for $ERDE_{50}$ was achieved by the German team **FHDO-BCSGD**, as well as the highest recall measure. Regarding the remaining metrics, the best F1 was obtained for the German team with the variant **FHDO-BCSGE**, while for precision, our team obtained the highest (0.91) using another method called Sequential Incremental Classification (SIC) (**UNSLD**) [7], although this value is now outperformed by the best precision reached with BEMB (1.0).

Although our UNSLB approach also used k-TVT, it is worth noting that the result is different from those shown in Subsect. 4.1 (see Table 2), since in the laboratory we adjusted the classifier parameters used (the logistic regressor) to obtain the best results for the competition. While in Table 2, on the contrary, we leave the parameters of each classifier by default.

---

[5] As it ran on a virtual platform Google Colab, memory and disk resources depend on the allocation of the moment, having only chosen GPU environment.

Table 3. Comparison with the best results in 2018 Lab for early detection of anorexia

	Classifier	p	$ERDE_\theta$		F1	Pre	Re
			$\theta = 5$	$\theta = 50$			
BOW	LR	0.6	11.52	8.68	0.62	0.78	0.51
5-TVT	RF	0.7	11.96	7.36	0.71	0.62	0.82
BEMB	NB	0.6	13.08	7.75	0.76	0.71	0.83
BEMB	RF	0.6	12.81	12.50	0.09	**1.00**	0.05
UNSLB	LR	0.6	**11.40**	7.82	0.61	0.75	0.51
UNSLD	–	0.6	12.93	9.85	0.79	0.91	0.71
FHDO-BCSGD	–	0.6	12.15	**5.96**	0.81	0.75	**0.88**
FHDO-BCSGE	–	0.6	11.98	6.61	**0.85**	0.87	0.83

Finally, we can conclude that although the analyzed approaches have not achieved the best performance obtained by the winners in the eRisk 2018 laboratory, their results are acceptable despite the generality of the models. Besides, if factors such as size of the text representations, use of resources and computational complexity are considered, k-TVT represents a suitable model to face this type of task where the most important thing is the time of delay in answering.

4.3 Performance of the Different Deep Learning Classification Methods

Before showing the results obtained, some considerations that were taken into account in the implementation will be mentioned.

The deep learning models were programmed in the Python language, in Google's Colab online platform and the pytorch library has been used for neural networks. In all implementations the Adam optimizer was used. In addition, due to the computing capacity of that platform, the complete vocabulary was not considered, but an approximate amount of 20,000 words was taken.

All the methods were tested configuring the parameters in different ways and the results shown correspond to the best configuration obtained by each method.

For the convolutional networks, embedding of 300 of GloVe was used as input of the method. Three types of filters 3, 5 and 7 were also used. A ReLU layer and also max-pooling were considered.

LSTM networks were also implemented with an embedding size of 300.

Then, BERTtokenizer was used for the input representation for BERT, and the pre-trained model was the BERT-base variant as mentioned in Sect. 2, which was taken from the Hugging Face repository[6].

---

[6] https://huggingface.co/bert-base-cased.

Table 4. Comparison of the best results of models based in deep learning for early detection of anorexia

Method	$ERDE_\theta$					F1	Pre	Re
	$\theta=5$	$\theta=10$	$\theta=25$	$\theta=50$	$\theta=75$			
CNN	**11.92**	**11.42**	**9.07**	**6.81**	**6.80**	**0.68**	**0.67**	**0.68**
LSTM	12.89	12.55	12.51	12.51	12.51	0.08	0.20	0.05
BERT	12.33	11.74	10.92	9.27	9.27	0.35	0.58	0.44

As can be seen, the convolutional neural network had a better performance compared to the other two methods. For the $ERDE_\theta$ measure, these networks performed very well when the level of urgency was lower ($\theta=50$ and $\theta=75$). Then, LSTM networks were the worst performers for both the $ERDE_\theta$ and the other measures.

Regarding the execution time, testing with 100 epochs, both CNN and LSTM took about 10 min. On the other hand, BERT, for 100 epochs, took several hours (approximately 7) and the results of the latter were not better than when it was run with less than 10 epochs (taking approximately 20 min).

4.4 Performance Comparison Between Models Using Standard Algorithms with Models Based on Deep Learning

Given the results displayed in the previous tables, we can see that regarding the relationship of the methods with the anorexia task, CNN obtained much better performance for the $ERDE_\theta$ measure when the level of urgency is lower ($\theta=50$ and $\theta=75$) and obtained the same performance that 5-TVT for $ERDE_{25}$. While for when the urgency is high ($\theta=5$) the Bag of Words is still better. Finally, none of the deep learning based methods outperformed the standard classifiers, for the F1, Precision and Recall measures.

It is also important to clarify that, although CNN obtained good results, it was not enough to exceed the values published in the eRisk2018 laboratory. However, this can be reversed if a more appropriate configuration is found for certain parameters that these deep learning models have.

5   Conclusions and Future Work

In this article, we focus on the task of early detection of anorexia disorder by evaluating the performance of different methods used in text mining. First, we analyze the text representations together with standard classifiers, and then we study models based on deep learning.

We can conclude in this work that, the approaches are competitive when it comes to solving the problem of early detection of anorexia, although they do not reach the values obtained in the eRisk2018 laboratory, since we consider that this could be improved with a better adjustment of parameters.

When urgency factor is priority, the best models include: Bag of Words (when the urgency is greatest), k-TVT, and convolutional networks. Whereas, when F1, Accuracy and Recall measures are considered, the model using BERT's embeddings was better.

Finally, k-TVT, although it did not obtain the best performance, is still a good option for tasks such as early risk detection, given its low computational complexity and dimensionality compared to the other text representations considered.

As future work, we want to modify the k-TVT method to process the data considering one writing at a time instead of taking a chunk by chunk (chunk approach) as we have been working on so far. In this way, we can use the data sets proposed in the latest editions of the eRisk labs. In addition, it is intended to carry out a study of the parameters in the methods based on deep learning to find a configuration that achieves a better performance.

References

1. Aguilera, J., Farías, D.I.H., Ortega-Mendoza, R.M., Montes-y-Gómez, M.: Depression and anorexia detection in social media as a one-class classification problem. Appl. Intell. **51**(8), 6088–6103 (2021). https://doi.org/10.1007/s10489-020-02131-2
2. Blumenstock, J.E.: Size matters: word count as a measure of quality on Wikipedia (2008)
3. Bucur, A.M., Cosma, A., Dinu, L.P.: Early risk detection of pathological gambling, self-harm and depression using BERT. In: CLEF (2021)
4. Cagnina, L., Errecalde, M.L., Garciarena Ucelay, M.J., Funez, D.G., Villegas, M.P.: k-TVT: a flexible and effective method for early depression detection. In: XXV Congreso Argentino de Ciencias de la Computación (CACIC) (2019)
5. Cavnar, W.B., Trenkle, J.M.: N-gram-based text categorization. In: Proceedings of SDAIR-94, 3rd Annual Symposium on Document Analysis and Information Retrieval (1994)
6. Devlin, J., Chang, M.W., Lee, K., Toutanova, K.: BERT: pre-training of deep bidirectional transformers for language understanding. ArXiv abs/1810.04805 (2019)
7. Funez, D.G., et al.: UNSL's participation at eRisk 2018 lab. In: CLEF (2018)
8. Harris, Z.S.: Distributional structure. Word **10**(2–3), 146–162 (1954)
9. Hochreiter, S., Schmidhuber, J.: Long short-term memory. Neural Comput. **9**(8), 1735–1780 (1997)
10. LeCun, Y., Bottou, L., Bengio, Y., Haffner, P.: Gradient-based learning applied to document recognition. Proc. IEEE **86**(11), 2278–2324 (1998)
11. McCulloch, W.S., Pitts, W.: A logical calculus of the ideas immanent in nervous activity. Bull. Math. Biophys. **5**(4), 115–133 (1943)
12. Mikolov, T., Sutskever, I., Chen, K., Corrado, G.S., Dean, J.: Distributed representations of words and phrases and their compositionality. In: NIPS (2013)
13. Pennington, J., Socher, R., Manning, C.D.: Glove: global vectors for word representation. In: EMNLP (2014)
14. Ramiandrisoa, F., Mothe, J.: Early detection of depression and anorexia from social media: a machine learning approach. In: CIRCLE (2020)

15. Rizwan, B., et al.: Increase in body dysmorphia and eating disorders among adolescents due to social media. Pakistan BioMed. J. **5**(1) (2022)
16. Rumelhart, D.E., Hinton, G.E., Williams, R.J.: Learning representations by back-propagating errors. Nature **323**(6088), 533–536 (1986)
17. Shah, F.M., et al.: Early depression detection from social network using deep learning techniques. In: 2020 IEEE Region 10 Symposium (TENSYMP) pp. 823–826 (2020)
18. Vaswani, A., et al.: Attention is all you need. ArXiv abs/1706.03762 (2017)
19. Villegas, M.P., Errecalde, M.L., Cagnina, L.: A comparison of text representation approaches for early detection of anorexia. In: XXVII Congreso Argentino de Ciencias de la Computación (CACIC) (Modalidad virtual), pp. 301–310 (2021)
20. Wang, Y.T., Huang, H.H., Chen, H.H.: A neural network approach to early risk detection of depression and anorexia on social media text. In: CLEF (2018)

GPU Permutation Index: Good Trade-Off Between Efficiency and Results Quality

Mariela Lopresti[ID], Fabiana Piccoli[(✉)][ID], and Nora Reyes[ID]

LIDIC, Universidad Nacional de San Luis,
Ejército de los Andes 950, 5700 San Luis, Argentina
{omlopres,mpiccoli,nreyes}@unsl.edu.ar

Abstract. When managing multimedia data such as text, images, videos, etc., it only makes sense to search for similar objects, because it is difficult to imagine that it would be interesting to look for if there is an element in the database exactly the same as another given as an example. Hence, a solution can be modeled through metric spaces. In this scenario, for solving efficient searches, we preprocess the database to build an index; and then, utilizing this index, minimize the number of comparisons required to answer them. However, with very large metric databases this is not enough, it is also necessary to speed up queries using high-performance computing. Then, the GPGPU appears as a profitable alternative. Moreover, there are circumstances in which it is also reasonable to accept quick answers even if they are inexact or approximate.

In this work, we evaluate the trade-off between the answer quality and performance of our GPU implementation of *Permutation Index*. The implementation is a pure GPU, used to solve in parallel multiple approximate similarity searches on metric databases. Our proposal has two parallelism levels: intra-queries and inter-queries, many queries are solved in parallel (inter-parallelism), and each query is figured out in parallel (intra-parallelism). The experimental results confirm that the *GPU Permutation Index* is a remarkable recourse to solve approximate similarity searches on large metric databases.

1 Introduction

When we work with multimedia objects: text, images, sound, etc., a search for a query object by equality in a database does not have sense. In this case, it is more meaningful to search for similar ones. Hence, we need to solve queries by measuring the similarity (or dissimilarity) between the query object and the objects in the database. The metric spaces allow us to model these similarity search problems. A metric space is defined by (U, d), where U is the universe of valid objects and d is the distance function or metric, which measures the similarity (or dissimilarity) among two given objects. d has to satisfy the positivity, symmetry, reflexivity, and triangular inequality properties. These properties make it a metric space. To solve a query on a database $X \subseteq U$, if $|X| = n$, the trivial method performs n evaluations of d between the query object and each element

P. Pesado and G. Gil (Eds.): CACIC 2021, CCIS 1584, pp. 183–200, 2022.
https://doi.org/10.1007/978-3-031-05903-2_13

of X. This solution is not good when n is very big. To reduce distance calculations, one way to solve it is through preprocessing X, i.e. creating an index. For a given query, an index helps to retrieve the most relevant objects from X by performing a few distance evaluations than n by each search [1]. One is the permutation index [2].

The *Permutation Index* is an approximate similarity search algorithm. It allows solving "inexact" similarity searches [3], prioritizing speed over accuracy or determinism in response [1,4]. There are many applications where their metric-space modelizations already involve an approximation to reality; hence, a second approximation at search time is usually acceptable.

When we work with very large metric databases, the index construction could not be enough to answer queries efficiently. Therefore other improvements are necessary. In this case, we include high-performance computing (HPC) techniques [5,6]. Graphics Processing Unit (GPU) [7] is a convenient tool to apply HPC in preprocessing stage of X and query solving. It is possible due to the characteristics of the GPU: it promises speedups of more than one order of magnitude compared to conventional processors for certain non-graphics calculations.

Particularly, in the case of metric spaces, several aspects accept optimization through HPC techniques, not only in query resolution but also in the preprocessing or index generation stage. Many GPU solutions for metric spaces exist. For example, to the k-nearest neighbors (k-NN) query has concentrated the most attention of researchers in this area, some are improvements to the brute force algorithm (sequential scanning or trivial solution) to find the k-NN of a given object [8–11], others for the creation of indexes [12,13] or to solve queries in them [14–16]. In [9] a *GPU Permutation Index* is proposed. They focus on high dimensional database and use Bitonic Sort to achieve good performance results.

This work has as objective to analyze the trade-off, when is used our *GPU Parallel Permutation Index*, between the answer quality of similarity queries and time performance. For the analysis, we consider different databases: type and sizes, response quality measures: Recall and Precision, and some performance parameters to evaluate HPC implementations.

This paper is an extended version of the paper accepted in CACIC 2021 [17], we include explanatory figures, extend the description and pseudo-codes of our approach, describe the experimental setup in more detail, add more results and elaborate further on the experimental discussion. It is organized as follows: the following sections describe theoretical concepts involved. Sections 4 and 5 develop the main characteristics of our proposal and its empirical performance. Finally, conclusions and future works are presented.

2   Metric Space Model

A metric space (U, d) is composed of a universe of valid objects U and a distance function $d : U \times U \rightarrow R^+$ defined among them. The distance function has to be particularly a metric, and determines the similarity (or dissimilarity) between two given objects: the smaller the distance, the closer or more similar are the objects. Hence, it satisfies the properties of a metric: positiveness ($\forall x, y \in U, d(x,y) \geq 0$), reflexivity ($\forall x \in U, d(x,x) = 0$), symmetry ($\forall x, y \in U, d(x,y) = d(y,x)$), and the triangle inequality ($\forall x, y, z \in U, d(x,z) \leq d(x,y) + d(y,z)$). Besides, in most cases d also satisfies the strict positiveness property ($\forall x, y \in U, x \neq y \rightarrow d(x,y) > 0$. The finite subset $X \subseteq U$ with size $n = |X|$, is called the *database* and represents the set of objects of the search space. It is usually to define the search complexity as the number of distance evaluations computed, disregarding other components because the distance is assumed to be expensive to calculate. There are two main kind of queries of interest [1,4]: Range queries and k-Nearest Neighbor queries (k-NN). The goal of a range search (q, r) is to retrieve all the objects $x \in X$ within the radius r of the query q (i.e. $(q,r) = \{x \in X / d(q,x) \leq r\}$). In k-NN queries, the objective is to retrieve the set k-NN$(q) \subseteq X$ such that $| \, k$-NN$(q) \, | = k$ and $\forall x \in k$-NN$(q), v \in X \wedge v \notin k$-NN$(q), d(q,x) \leq d(q,v)$.

A way to accelerate the searches is to preprocess the database X to build an index. The index helps answer queries with less than n distance evaluations to retrieve all the relevant objects from X. Each possible index saves different information obtained during the preprocessing of the database. Some indexes store a subset of distances between objects; others maintain only a range of distance values. In general, there is a trade-off between the quantity of information stored in the index and the query cost it achieves. As more information needs to save an index (more memory it uses), the lower query cost it could obtain. However, some indexes take better advantage of the stored information and memory used than others. In a database of n objects, it is possible to calculate $n(n-1)/2$ distances among all element pairs from the database X. Nevertheless, most of the indexes usually avoid storing this amount of information because $O(n^2)$ space is unacceptable for real applications [18].

Therefore, similarity searching in metric spaces usually is solved in two stages: preprocessing and query time. During the preprocessing stage, the index is built. Then, during the query time, the index is used to avoid some distance computations. The state-of-the-art in this area can be divided into two families [1]: *pivot-based algorithms* and *compact partition-based algorithms*.

Some indexes obtain the "exact" answer to the query; that is, all the relevant objects from the database. However, there is an alternative to "exact" similarity searching called *approximate similarity searching* [3], where accuracy or determinism is traded for faster searches [1,4]. This alternative encompasses *approximate* and *probabilistic algorithms*. The goal of approximate similarity search is to reduce *significantly* search times by allowing some "errors" in the query output. In approximate algorithms, a threshold ϵ is considered as a parameter. So, in a range query (q, r) the retrieved elements are guaranteed to be at most at distance $(1 + \epsilon)r$ to the query q [19]. This relaxation gives faster algorithms as the threshold ϵ increases [19,20]. On the other hand, probabilistic algorithms state that the answer is correct with high probability [21,22].

If a k-NN query of an element $q \in U$ is posed, the index answers with the k closest elements from X viewed, but only between the elements that are effectively compared with q. However, as we want to save as many distance evaluations as possible, q will not be compared with many potentially relevant elements. If the exact answer of k-NN$(q) = \{x_1, x_2, \ldots, x_k\}$, it determines the radius $r_k = \max_{1 \leq i \leq k}\{d(x_i, q)\}$ needed to enclose these k closest elements to q. Hence, the approximate k-NN search will obtain k elements too, but the needed radius to enclose this answer could be greater than r_k; that is, some closest relevant elements might not appear in the response.

Among all the indexes for approximate similarity searches, the Permutation-Based Algorithm (PBA) stands out for achieving good performance for approximate similarity queries [2]. Therefore, we consider the PBA index as the basis of our proposal. As it is well-known, we must not neglect the quality of the answer obtained in approximate similarity searches. Therefore, we evaluate our proposal considering both aspects: its answer quality and the number of distances calculated at searches. Hence, we introduce the description of the sequential PBA and the quality measures to be considered.

2.1   Sequential Permutation Index

Let $\mathcal{P}$ be a permutants set where $\mathcal{P} = \{p_1, p_2, \ldots, p_m\} \subseteq X$, i.e. each p_i is a database object of X. For each element $x \in X$, a vector is defined, called permutation of x (Π_x). It is formed by ordered distances of x to each permutant, $\Pi_x = \langle p_{i_1}, p_{i_2}, \ldots p_{i_m}\rangle$. Formally, for an element $x \in X$, its permutation Π_x of $\mathcal{P}$ satisfies $d(x, \Pi_x(i)) \leq d(x, \Pi_x(i+1))$. If two elements are the same distance, an arbitrary and consistent order is considered. We use $\Pi_x^{-1}(p_{i_j})$ for the *rank* of an element p_{i_j} in the permutation Π_x. The index works on a straightforward principle: two similar elements x and y would have similar permutations; that is, if x is similar to y, Π_x would be similar to Π_y [2].

The *Permutation-based Algorithm* (PBA) is a probabilistic algorithm example, the elements similarity is predicted through permutations. Its query algorithm is very simple, it consists of the following steps:

1. Pre-processing Stage: $\forall x \in X$, their permutations are calculated. All permutations integrate the permutation index.
2. Query Stage: For a q query:
   (a) Its permutation Π_q is calculated.
   (b) The similarity between Π_q and all elements in X is computed, ordering them by ascending values of "dissimilarity" between permutations.
   (c) Finally, q is compared using d against the first objects in determined list by the previous step. This is done until a stop criterion is reached and depending on the type of query to be performed.

There are several ways to measure the (dis)similarity between two permutations, they can be, for example, *Kendall Tau, Spearman Rho*, or *Spearman Footrule* metrics [23]. All these measures satisfy the properties required to be a metric. Between them, the Spearman Footrule metric is used because it is not expensive to calculate, and it predicts very well the proximity or similarity between elements [2]. The Spearman Footrule distance is the *Manhattan distance* L_1 between permutations, it belongs to the family of Minkowsky distances. Formally, the Spearman Footrule metric F is defined as:

$$F(\Pi_x, \Pi_q) = \sum_{i=1}^{m} |\Pi_x^{-1}(p_i) - \Pi_q^{-1}(p_i)|$$

As the previous query algorithm describes, after finishing the first two steps, we traverse X according to the order obtained by evaluating the distance $d(q, x)$ for each $x \in X$. For range queries, with radius r, each x that satisfies $d(q, x) \leq r$ is reported, and for k-NN query, the set of k elements with smaller distance. Only a fraction $f\%$ of the database is visited, the remaining elements are ignored. This is the reason that makes probabilistic the algorithm, even if $F(\Pi_q, \Pi_x) < F(\Pi_q, \Pi_v)$ does not guarantee that $d(q, x) < d(q, v)$, being able stop the search prematurely. When the order based on $F(\Pi_q, \Pi_x)$ is close to real order of $d(q, u)$, the algorithm works very well. The efficiency and quality of answers obviously depend on the parameter f. In [2], good f values are discussed.

The Fig. 1 depicts an example of objects and a query object q in $\mathbb{R}^2$, using Euclidean distance, and a permutant set $\mathcal{P} = \{p_1, p_2, p_3, p_4, p_5, p_6\}$. The permutation of each database object and of q are shown. It can be noticed that the most similar object to q is u_1 and their permutations are also similar.

2.2   Quality Measures of Approximate Search

As we mention previously, an approximate answer of k-NN(q) could obtain some elements z whose $d(q, z) > r_k$. Besides, an approximate range query of (q, r) could obtain a subset of the exact answer because the algorithm possibly did not have reviewed all the relevant elements. However, the distance between each object in the answer set will be less or equal to r, so they also belong to the exact answer of (q, r).

In most of the Information Retrieval (IR) systems, it is necessary to evaluate the retrieval effectiveness [24]. There are many measures of retrieval effectiveness proposed. In IR systems, the most commonly used are called *Recall* and *Precision*. We call as $Retr(q)$ the set of retrieved objects for a query q in an approximate search, and as $Rel(q)$ the set of all relevant elements in the database X

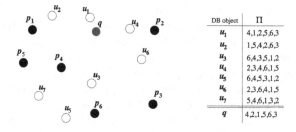

DB object	Π
u_1	4,1,2,5,6,3
u_2	1,5,4,2,6,3
u_3	6,4,3,5,1,2
u_4	2,3,4,6,1,5
u_5	6,4,5,3,1,2
u_6	2,3,6,4,1,5
u_7	5,4,6,1,3,2
q	4,2,1,5,6,3

Fig. 1. Example of a database in $\mathbb{R}^2$ and its permutations.

for the query q. *Recall* is the ratio of the *number of relevant elements retrieved* for a given query ($|Retr(q) \cap Rel(q)|$) over the *number of relevant elements* for this query in the database ($|Rel(q)|$). *Precision* is the ratio of the *number of relevant elements retrieved* ($|Retr(q) \cap Rel(q)|$) over the *total number of elements retrieved* ($|Retr(q)|$). Both quality measures, Recall and Precision, take on their values between 0 and 1. In IR systems the elements are documents.

In small text collections on general IR systems, the denominator of both ratios is generally unknown and must be estimated by sampling or some other method. In our scenario, we could obtain the exact answer for each query q, as the set of relevant elements for this query in X. In this way, it is possible to evaluate both measures for an approximate similarity search index.

For each query element q, the exact k-NN(q) = $Rel(q)$ is determined with some exact metric access method. The approximate-k-NN(q) = $Retr(q)$ is answered with an approximate similarity search index. Let be the set $Retr(q) = \{y_1, y_2, \ldots, y_k\}$. It can be noticed that the approximate search will also return k elements, so $|Retr(q)| = |Rel(q)| = k$. Thus, we can establish the number of k elements obtained which are relevant to q by verifying if $d(q, y_i) \leq r_k$; that is, calculating $|Rel(q) \cap Retr(q)|$. Hence, on approximate k-NN both measures obtain the same result and will allow us to verify the effectiveness of our proposal:

$$Recall = \frac{|Rel(q) \cap Retr(q)|}{|Rel(q)|} = \frac{|Rel(q) \cap Retr(q)|}{k}$$

and

$$Precision = \frac{|Rel(q) \cap Retr(q)|}{|Retr(q)|} = \frac{|Rel(q) \cap Retr(q)|}{k},$$

In approximate range queries the Precision is always equal to 1 because $Retr(q) \cap Rel(q) = Retr(q)$; that is:

$$Precision = \frac{|Rel(q) \cap Retr(q)|}{|Retr(q)|} = \frac{|Retr(q)|}{|Retr(q)|} = 1,$$

Thus, we only consider the Recall measure to analyze the retrieval effectiveness (answer quality) of our proposal for both kinds of queries (k-NN and range queries).

2.3   GPGPU Programming

GPGPU (General-Purpose GPU) is to use the GPU to solve general-purpose computing, not just computing of a graphical nature [7,25,26]. The parallel programming over GPUs has its own characteristics, all of these are differences respect typical parallel programming (in parallel computer), the most relevant are: the number of processing units, the CPU-GPU memory structure, and the number of parallel threads. Not all problems types can be solved in a GPU architecture, the suitable problems are those that can be implemented with stream processing and using limited memory, that is, applications with abundant parallelism. Each GPGPU algorithm must be carefully analyzed and data structures must be designed considering its memory hierarchy, architecture and limitations.

Every GPGPU program has many basic steps, at the beginning a single CPU process starts that has to transfer the input data to GPU memory. Once the data are in place on the card, it launches hundreds of threads on GPU (with little overhead). Each thread works over its own data and, at the end of the computation, the results should be copied back to CPU main memory. This is the natural form to generate work on GPU: All of them share the same space memory, and they are able to execute independently and at the same time. The traditional multi-threading is used to do time-slicing or take advantage of idle time, i.e. while a thread waits, another could execute. In consequence, a good GPGPU algorithm has to take account:

- Overlapping or reducing data transfers between the CPU and the GPU, they are significantly time-consuming.
- The algorithm must adopt to the MIMD and SIMD paradigms, and accept the SIMT execution model.
- A lot of work amount has to be generated to efficiently use all GPU cores.

The Compute Unified Device Architecture (CUDA) allows using the GPU as a highly parallel computer for general-purpose computing [7,27]. It was developed by NVidia to own GPU and provides an essential high- Level development environment with standard high level programming language (C, C++, Fortran, Python). It considers the GPU as a CPU co-processor and defines its architecture and a programming model: parallel-concurrent threads and the memory hierarchy. A CUDA program consists of multiple phases that run on CPU (sequential code) or GPU (kernel, parallel code). A kernel function defines that to be executed by each thread with its data on GPU.

3   GPU-Permutation Index

The two processes of Permutation Index: indexing and query resolution, can be solve in GPU. As the sequential process, first the index has to be created, and next all queries are answered. In GPU, the Index building process has two stages and the query process, four (as it is described previously). The Fig. 2 shows the architecture of the general system.

Building a permutation index in GPU involves two steps, a pseudo-code is showed in Algorithm 1. The first one is $Distances(x, \mathcal{P})$, which calculates the distance among every object $x \in X$ and each $p \in \mathcal{P}$ (line 3). The second one, $Permutation\ (x, \mathbb{S})$, sets up the signatures $\forall x \in X$, line 4–5, to add them to the Permutation Index $\mathbb{I}$. The process input is X and $\mathcal{P}$. The output is the index $\mathbb{I}$ ready to be queried. The idea is to divide the work into threads blocks; in parallel (line 2), each thread calculates the object permutation according to a global $\mathcal{P}$, line 1.

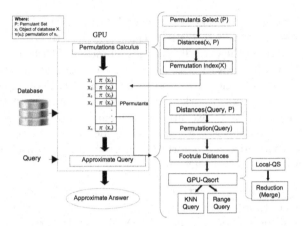

Fig. 2. Indexing and querying in GPU-CUDA Permutation Index.

Algorithm 1. Permutation Index

1: $\mathbb{P} \leftarrow$ Select Permutants $(\mathbb{X})$, $\mathbb{I} \leftarrow \emptyset$ ▷ $\mathbb{X}$:Database, $\mathbb{I}$: Index
2: **for all** $x \in \mathbb{X}$ **do**
3:      $\mathbb{D} \leftarrow$ Distances$(x, \mathbb{P})$
4:      $\mathbb{S} \leftarrow$ Sort $(\mathbb{D})$
5:      $\Pi_x \leftarrow$ Permutation$(x, \mathbb{S})$
6:      $\mathbb{I} \leftarrow \mathbb{I} \cup \{\Pi_x\}$ ▷ Get index search in $\mathbb{I}$
7: **end for**

In $Distances(x, \mathcal{P})$, the number of blocks will be defined according of X size and the threads number per block which depends of required resources quantity by block. Finally, each threads block saves in the device memory its calculated distances. This stage requires a size structure $m \times n$ (m: permutants number, and n: $|X|$), and an auxiliary structure in the Shared Memory (it stores the permutants, the process is repeated when the permutants size is greater than auxiliary structure size). Each thread calculates one object permutation: it takes one object and sorts the permutants according to their distance. The output of second step is the *Permutation Index* $\mathbb{I}$ and needs a $n \times m$ size of the device memory. The *Permutation Index(X)* would answer different kind of approximated

---

Algorithm 2. Query Solution for q

1: $\mathbb{D} \leftarrow$ Distances $(q, \mathbb{P})$
2: $\mathbb{S} \leftarrow$ Sort $(\mathbb{D})$
3: $\Pi_q \leftarrow$ Permutation $(q, \mathbb{S})$
4: $\mathbb{F} \leftarrow$ Footrule Distances$(\Pi_q, \mathbb{I}))$ ▷ Footrule distances between Π_q and perms. in $\mathbb{I}$
5: $\mathbb{O} \leftarrow Qsort(\mathbb{I})$ ▷ Sort the elements of $\mathbb{I}$ using F
6: opción $\leftarrow$ *Search by k-NN or Range*
7: **if** opción $= k$-NN **then** k-NN Search $(q, \mathbb{O})$
8: **else** *Range Search* $(q, \mathbb{O})$
9: **end if**

---

queries, the most common are *by range* or *k*-NN. *Query process* implies four steps. In Algorithm 2, we show a simplify pseudo-code. First, the query object permutation is computed in parallel by so many threads as permutants exist, lines 1–3. The next step, Footrule Distance($\Pi_q, \mathbb{I}$)), is to compare all permutations in the index $\mathbb{I}$ with the query permutation. By each thread, the *Footrule* distances are calculated between the query permutation and the permutation of one object in X, line 4. In the third step, all objects in $\mathbb{I}$ are sorted by their distances F (line 5). Finally, in lines 6–9, depending of query kind, the candidate objects have to be evaluated by means of the *real distance* with query object again. As we say before, a database percentage is considered for this step, for example the 10% (it can be a parameter).

As sorting method, we develop the Quick-sort in the GPU, GPU-Qsort, considering the highly parallel nature of GPUs. Its main characteristics are: iterative algorithm and a good use of Shared memory [28].

The GPU characteristics and developed GPU-Permutation Index us allow to solve many approximated queries in parallel. The Fig. 3 presents the few needed changes of GPU-Permutation Index to solve many queries at the same time. The difference is in *Query process*, because the Permutation Index is built once and then it is used to answer all queries. For *Query process* answers in parallel many approximate queries. How does it work? A queries set is sent to GPU and, by each query in parallel, applies the process explained before, Fig. 2. The needed resources size is equal to resources amount of one query multiplied by the solved queries number in parallel. The solution implies a care management of blocks and their threads. Therefore, it is important a good administration of threads. Each thread has to know which query it is solving and which database object. This is possible by establishing a relationship among *Thread Id*, *Block Id*, *Query Id*, and *Database Element*.

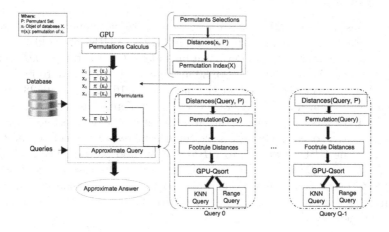

Fig. 3. Solving many queries in GPU-CUDA Permutation Index.

The solved parallel queries number is determined depending on the GPU resources, principally its memory: Global and Shared. The total required memory to solve parallel queries is $Q * m + i$, where Q is parallel queries number, m the needed memory quantity per query and i the needed memory by the Permutation Index. Once Q parallel queries are answered, the results can be sent to CPU or wait until all results of each subset of Q queries, and transfer them all together via PCI-Express.

4   Experimental Results

In this section, we present the experimental results of GPU permutation index: Performance and answers Quality. The experiments were developed in different scenarios. For the sequential versions a computer has the following characteristics: Intel(R) Xeon(R) processor CPU E5-2603 v2 @ 1.80 GHz x 8–15.6 GB memory. For parallel versions, different GPUs are considered which belong to different architecture generations or have different amount of resources. Their characteristics are (GPU Model, Memory, CUDA Cores, Clock Rate and Capability):

– Tesla K20c, 4800 MB, 2496, 0.71 GHz, 3.5.
– GTX470, 1216 MB, 448, 1.22 GHz, 2.0.

Respect of the databases, two of them were selected from SISAP METRIC SPACE LIBRARY (www.sisap.org). Their characteristics are:

– English words (DBs) and using the *Levenshtein* distance, also called *edit* distance.
– Colors histograms (DBh) is formed by 112-vectors. In this case, we use Euclidean distance.

In both cases, different DB size are considered. So, we named each subset as DBs or DBh adding a suffix with its size in kB (+ <DBsize> in kB).

For the parameter f of the Permutation Index (f = fraction of DB revised during searches). We consider 10, 20, 30 and 50% of the DB size to be revised. The number of permutants used for the index are $5, 16, 32, 64$, and 128. In each case the results shown are the average over 1000 different queries.

In Fig. 4, we show the Index Creation times for all the devices and databases. For all devices and each DB, the performance increases respect to the CPU. Considering the GPU resources, to DBh, a very good performance is obtained in the $DBh4$ and $DBh97$ cases for both devices. In the case of the $DBh500$, the GTX 470 time is not shown due to not having enough memory to solve the problem (this is a future work line: dealing with databases size larger or greater than GPU memory).

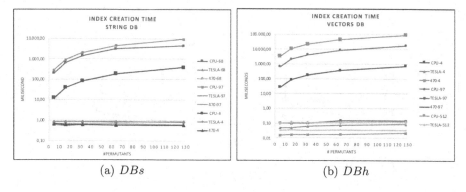

(a) *DBs*   (b) *DBh*

Fig. 4. Time of index creation for both databases.

In this case, only the times to build the index w ere taken into account in the comparisons, but we measure the transfer time of the complete *DB*. In our case, both devices have the same PCI Express technology. For example the transfer time for:

- *DBs97* is 1.23 ms. It consumes 60% of the total process time in the Tesla K20c and 66% in the GTX 470.
- *DBh500* is 69.65 ms. It implies the 99% of the total process time in the Tesla K20c.

In Figs. 5 and 6, we show the obtained time in k-NN and range queries respectively, for different parameters: permutants number, range, k, and *DB* percentage. The k values are 3 and 5; and the radii range depend of databases, for *DBs* are 1, 2, and 3, and *DBh* are 0,05, 0,08 and 0,13. In these results, 80 queries are solved in parallel. As it can be noticed *Range queries* show improvements respect to k-NN queries, but in both cases the achieved times are much less than CPU times. In all cases, it is clear the influence of *DB* size, but evenly we accomplish good performance. In all cases, the permutants number does not influence the time.

In Fig. 7, we can see how the queries number to be solved in parallel influences the performance. Shorter times are achieved when queries number is greater. For *DBh*, the time to solve 1 (one) query vs 30 (thirty) queries decreases in the best of cases in the order of 1.8x (Tp1/Tp30). In the case of the *DBs*, the gains obtained in solving multiple queries in parallel are greater: an improvement of 2.5x.

For the case of k-NN, the times are similar. This behavior is similar in both *DB*, i.e. the GPU resources have more work to do and, consequently, less idle time.

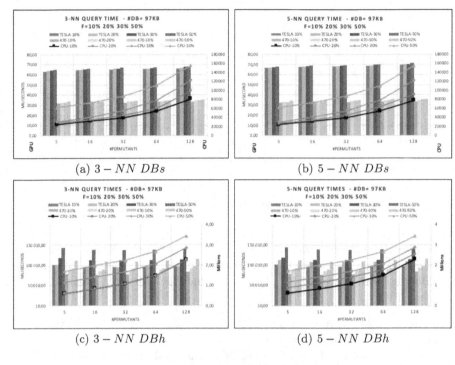

(a) $3 - NN\ DBs$ (b) $5 - NN\ DBs$

(c) $3 - NN\ DBh$ (d) $5 - NN\ DBh$

Fig. 5. k-NN query time for both databases.

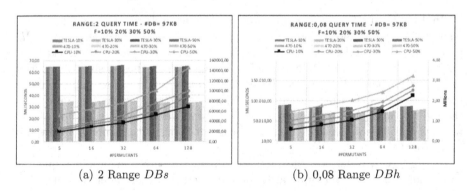

(a) 2 Range DBs (b) 0,08 Range DBh

Fig. 6. Range query time for both databases.

Other evaluated aspect was permutants number influence in each query kind solved in GPU. In Fig. 8, we display the results for different: permutants number, DB, queries type, and DB fraction (10% and 50%). For this and each DB, we consider different permutants, generate the Permutation Index and solve

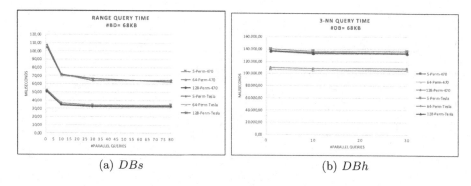

(a) *DBs*   (b) *DBh*

Fig. 7. Multi-queries time for both databases.

the queries. The permutants number determine required memory quantity. The Figs. 4a and 4b show a comparison for *DBs*97 when we vary the permutants number. A minimal times variation can be observed for both GPUs, it does not reach 2 ms. The same behavior is observed to *DBh*, the permutants number does not affect time. Figures 4a and 4b show this for both types of queries.

If we increase permutants number, greater improvements are obtained because the sequential solution demands more time. For the other side, while the parallel solution in any GPU remains almost constant in time. The same behavior is replicated in the other sizes of *DBs*.

The trade-off between the answer quality and time performance of our parallel index with respect to the sequential index is good. To evaluate answer quality of k-NN or range query, it is necessary to obtain the exact answers, $Rel()$. It is used to determine quality of the approximate answer, $Retr()$. For the evaluation, we consider both queries kinds, the Permutation Index respectively with 5, 64 and 128 permutants, and different *DB* percentages. Figures 9 and 10 illustrate the average answer quality obtained. As it can be noticed, the GPU Permutation Index retrieves a good percentage of exact answer only reviewing a little fraction of the *DB*. For example, the 10% retrieves 40% and it needs to review the 30% to retrieve almost 80% of exact answer.

We show only a representative subset of database sizes and permutants numbers, the other sizes and numbers have yielded similar results.

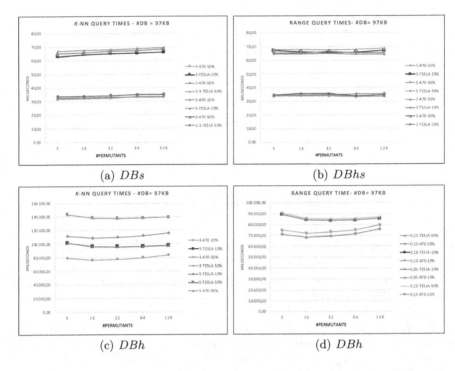

(a) *DBs*   (b) *DBhs*

(c) *DBh*   (d) *DBh*

Fig. 8. Influence of the number of permutants in GPU Permutation Index

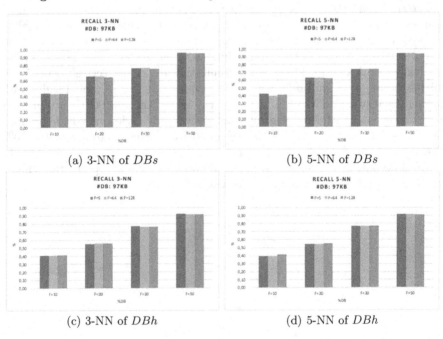

(a) 3-NN of *DBs*   (b) 5-NN of *DBs*

(c) 3-NN of *DBh*   (d) 5-NN of *DBh*

Fig. 9. Recall of approximate-*k*-NN queries for both databases.

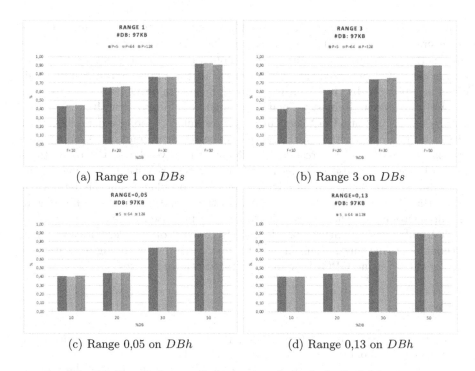

(a) Range 1 on DBs (b) Range 3 on DBs

(c) Range 0,05 on DBh (d) Range 0,13 on DBh

Fig. 10. Recall of approximate-range queries for both databases.

5 Conclusions and Future Works

The application of high-performance techniques to solve general-purpose problems on GPU enables us to obtain better solutions. Therefore, in this work, we consider three fundamental aspects: HPC in GPU, Metric Spaces, and their relationship; that is, we consider how to obtain better solutions in metric spaces by applying GPGPU. For this objective, we select the Permutation Index, a well-known index for approximate similarity search on metric space, and implement indexing and query resolution processes on GPU. Moreover, we develop our solution with two parallelism levels: intra-queries and inter-queries. This new implementation is named GPU Permutation Index.

In this paper, we show the main characteristics of GPU Permutation Index and evaluate its performance to accomplish approximate similarity searches on metric databases and the quality of the answers obtained.

For the experimental evaluation, we consider two metric databases: one consists of English words and the other of vectors of color histograms. The results obtained exhibits improvements achieved on the different GPUs. The speedup and performance of our proposal are significant, showing that the enhancements grow as workload increases. From the other point of view, regarding answer quality, the results are analyzed using the *Recall* metric and also validate the proposed GPU solution.

Other achievements, besides the outstanding performance obtained to one or several queries and answer quality, is that our proposal allows increasing the fraction f of database to examine without the risk of affecting performance significantly and achieving greater accuracy in the approximate results. This claim is supported by the extensive validation process developed.

As future research lines, we plan to evaluate our proposal with other metric databases (documents, DNA sequences, images, music, among others) and other distance functions. In this work, we select the set of permutants randomly, so we plan to study how the results are affected when we choose the permutants in another way. We also consider analyzing different partition strategies of the databases and/or using several GPUs to solve the problem when the database is larger than the size of the GPU memory. In this situation, we could investigate the application of dynamic parallelism in the proposed solution, considering the Permutations Signatures [29] to reduce the size of the Permutation Index without removing any permutant. Furthermore, we regard taking advantage of the benefits of using GPGPU with other metric space indexes.

References

1. Chávez, E., Navarro, G., Baeza-Yates, R., Marroquín, J.: Searching in metric spaces. ACM Comput. Surv. **33**(3), 273–321 (2001)
2. Chávez, E., Figueroa, K., Navarro, G.: Proximity searching in high dimensional spaces with a proximity preserving order. In: Gelbukh, A., de Albornoz, Á., Terashima-Marín, H. (eds.) MICAI 2005. LNCS (LNAI), vol. 3789, pp. 405–414. Springer, Heidelberg (2005). https://doi.org/10.1007/11579427_41
3. Ciaccia, P., Patella, M.: Approximate and probabilistic methods. SIGSPATIAL Spec. **2**(2), 16–19 (2010). https://doi.org/10.1145/1862413.1862418
4. Zezula, P., Amato, G., Dohnal, V., Batko, M.: Similarity Search: The Metric Space Approach, ser. Advances in Database Systems, vol. 32. Springer, Boston (2006). https://doi.org/10.1007/0-387-29151-2
5. Pacheco, P., Malensek, M.: An Introduction to Parallel Programming, ser. An Introduction to Parallel Programming. Elsevier Science (2021). https://books.google.com.ar/books?id=uAfXnQAACAAJ
6. Robey, R., Zamora, Y.: Parallel and High Performance Computing. Simon and Schuster (2021). https://books.google.com.ar/books?id=jNstEAAAQBAJ
7. Kirk, D.B., Hwu, W.: Programming Massively Parallel Processors, A Hands on Approach. Elsevier, Morgan Kaufmann (2017). https://doi.org/10.1016/C2015-0-02431-5
8. Barrientos, R., Millaguir, F., Sánchez, J.L., Arias, E.: GPU-based exhaustive algorithms processing KNN queries. J. Supercomput. **73**, 4611–4634 (2017)
9. Kruliš, M., Osipyan, H., Marchand-Maillet, S.: Employing GPU architectures for permutation-based indexing. Multimedia Tools Appl. **76**(05) (2017). https://doi.org/10.1007/s11042-016-3677-7
10. Li, S., Amenta, N.: Brute-Force k-nearest neighbors search on the GPU. In: Amato, G., Connor, R., Falchi, F., Gennaro, C. (eds.) SISAP 2015. LNCS, vol. 9371, pp. 259–270. Springer, Cham (2015). https://doi.org/10.1007/978-3-319-25087-8_25

11. Velentzas, P., Vassilakopoulos, M., Corral, A.: In-memory k nearest neighbor GPU-based query processing. In: Proceedings of the 6th International Conference on Geographical Information Systems Theory, Applications and Management - GIS-TAM, INSTICC, pp. 310–317. SciTePress (2020)

12. Barrientos, R.J., Gómez, J.I., Tenllado, C., Prieto, M., Zezula, P.: Multi-level clustering on metric spaces using a multi-GPU platform. In: Euro-Par (2013)

13. Eder dos Santos, R.U.P., Sofia, A.A.O.: Procesamiento de búsquedas por similitud. tecnologías de paralelización e indexación. In: Informe Científico Técnico UNPA, vol. 7, no. 2, pp. 111–138 (2015)

14. Gowanlock, M., Karsin, B.: Accelerating the similarity self-join using the GPU. J. Parallel Distrib. Comput. **133**, 06 (2019)

15. Barrientos, R.J., Riquelme, J.A., Hernández-García, R., Navarro, C.A., Soto-Silva, W.: Fast KNN query processing over a multi-node GPU environment. J. Supercomput., 3045–3071 (2022)

16. Riquelme, J.A., Barrientos, R.J., Hernández-García, R., Navarro, C.A.: An exhaustive algorithm based on GPU to process a KNN query. In: 2020 39th International Conference of the Chilean Computer Science Society (SCCC), pp. 1–8 (2020)

17. Lopresti, M., Piccoli, F., Reyes, N.: Goodness of the GPU permutation index: performance and quality results. In: XXVII Congreso Argentino de Ciencias de la Computación, CACIC 2021, pp. 321–332 (2021)

18. Figueroa, K., Chávez, E., Navarro, G., Paredes, R.: Speeding up spatial approximation search in metric spaces. ACM J. Exp. Algorithmics **14** (2009). Article 3.6

19. Bustos, B., Navarro, G.: Probabilistic proximity searching algorithms based on compact partitions. J. Discrete Algorithms **2**(1), 115–134 (2004). The 9th International Symposium on String Processing and Information Retrieval. https://www.sciencedirect.com/science/article/pii/S1570866703000674

20. Tokoro, K., Yamaguchi, K., Masuda, S.: Improvements of TLAESA nearest neighbour search algorithm and extension to approximation search. In: Proceedings of the 29th Australasian Computer Science Conference - Volume 48, ser. ACSC 2006, Darlinghurst, Australia, pp. 77–83. Australian Computer Society Inc., Australia (2006). http://dl.acm.org/citation.cfm?id=1151699.1151709

21. Singh, A., Ferhatosmanoglu, H., Tosun, A.: High dimensional reverse nearest neighbor queries. In: The Twelfth International Conference on Information and Knowledge Management, ser. CIKM 2003, pp. 91–98. ACM, New York (2003). https://doi.org/10.1145/956863.956882

22. Moreno-Seco, F., Micó, L., Oncina, J.: A modification of the LAESA algorithm for approximated k-NN classification. Pattern Recogn. Lett. **24**(1–3), 47–53 (2003). http://www.sciencedirect.com/science/article/pii/S0167865502001873

23. Fagin, R., Kumar, R., Sivakumar, D.: Comparing top k lists. In: Proceedings of the Fourteenth Annual ACM-SIAM Symposium on Discrete Algorithms, ser. SODA 2003, pp. 28–36. Society for Industrial and Applied Mathematics, Philadelphia (2003). http://dl.acm.org/citation.cfm?id=644108.644113

24. Baeza-Yates, R.A., Ribeiro-Neto, B.A.: Modern Information Retrieval. Pearson Education Ltd., Harlow (2011)

25. Cheng, J., Grossman, M.: Professional CUDA C Programming. CreateSpace Independent Publishing Platform (2017). https://books.google.com.ar/books?id=ApBtswEACAAJ

26. Han, J., Sharma, B.: Learn CUDA Programming: A Beginner's Guide to GPU Programming and Parallel Computing with CUDA 10.x and C/C++. Packt Publishing (2019). https://books.google.com.ar/books?id=dhWzDwAAQBAJ

27. NVIDIA: Nvidia CUDA compute unified device architecture, programming guide, in NVIDIA (2020)
28. Lopresti, M., Miranda, N., Piccoli, F., Reyes, N.: Permutation index and GPU to solve efficiently many queries. In: VI Latin American Symposium on High Performance Computing, HPCLatAm 2013, pp. 101–112 (2013)
29. Figueroa, K., Reyes, N.: Permutation's signatures for proximity searching in metric spaces. In: Amato, G., Gennaro, C., Oria, V., Radovanović, M. (eds.) SISAP 2019. LNCS, vol. 11807, pp. 151–159. Springer, Cham (2019). https://doi.org/10.1007/978-3-030-32047-8_14

A Comparison of DBMSs for Mobile Devices

Fernando Tesone[(✉)][iD], Pablo Thomas[iD], Luciano Marrero[iD], Verena Olsowy[iD], and Patricia Pesado[iD]

Instituto de Investigación en Informática LIDI, Facultad de Informática,
Universidad Nacional de La Plata, La Plata, Argentina
{ftesone,pthomas,lmarrero,volsowy,ppesado}@lidi.info.unlp.edu.ar

Abstract. With the growth in the reach and use of the Internet, smartphones and social networks, there is an exponential increase in the volume of data managed, which can be structured, semi-structured or unstructured. In this context, NoSQL databases are emerging, which facilitate the storage of semi-structured or unstructured data. In this context, NoSQL databases emerge, which facilitate the storage of semi-structured or unstructured data.

On the other hand, improvements in hardware performance of mobile devices are leading to more information being managed by these, and to the emergence of new database management systems that are installed on these devices. The purpose of this work is to survey the database management systems for mobile devices, and to perform an experimental analysis of the most representative systems of the most widely used database models.

Keywords: Databases for mobile devices · Relational DBMS · NoSQL DBMS · Document store DBMS

1 Introduction

Database management systems (DBMS) have played a key role in software development since their emergence in the 1960s, providing an efficient way to generate complex applications by eliminating the need for programming persistence and data access [1,2].

In 1970 Edgar Codd developed the relational database model, which has become the dominant model since then [1,3].

The *Apple iPhone* was introduced in 2007 and the *Android* operating system in 2008, events that would radically change the smartphone industry, as they increased the popularity of the use of mobile devices, reaching 80% of the population in some countries [4–6]. This growth in usage would bring with it a diversification of platforms. To maximize market presence, *apps* must be available on multiple platforms or operating systems, so software developers must opt for native, platform-specific, or cross-platform developments [7,8].

P. Pesado and G. Gil (Eds.): CACIC 2021, CCIS 1584, pp. 201–215, 2022.
https://doi.org/10.1007/978-3-031-05903-2_14

With the growth in the reach and use of the Internet and mobile devices, coupled with the emergence of social networks, there is an exponential growth in the volume of data managed [9], which can be structured, semi-structured or unstructured. Faced with this situation of growth in the volume of information, non-relational or NoSQL (Not-only SQL) databases have emerged as an alternative to relational databases, which facilitate the massive storage of semi-structured or unstructured data.

With the emergence of these technologies, software developers must analyze which DBMSs are suitable for the needs of the problem to be solved.

The objective of this work is (1) to carry out a survey of existing DBMSs, both relational and non-relational, for mobile devices —i.e. that can be embedded in applications for mobile devices—, (2) to select DBMSs that are considered representative of the surveyed set, based on the definition of a series of criteria, and (3) to carry out an experiment that allows to analyze, for each selected DBMS, specific characteristics, advantages and disadvantages, from the point of view of the software engineer's experience.

This paper represents an extension of the publication presented at the *Congreso Argentino de Ciencias de la Computación* (CACIC) in its 2021 edition [10]. In relation to the previous work, this paper modifies the defined DBMS selection criteria, thus including Firebase Realtime Database in the performed experimentation and subsequent analysis. Furthermore, the analysis of the results obtained from the experimentation and the conclusion presented here, were enriched by the inclusion of the aforementioned DBMS.

This work is organized as follows: related work is discussed in Sect. 2; a survey of database management systems for mobile devices is presented in Sect. 3; in Sect. 4, representative DBMSs are selected, and experimentation is carried out to analyze specific characteristics, and advantages and disadvantages of each selected DBMS. Subsequently, in Sect. 5 the results obtained from the experimentation carried out are analysed. Finally, in Sect. 6 the conclusions are presented and possible lines of research for future work are defined.

2   Related Work

This section describes the related works found, linked to the topic presented.

In [11], a list of features that mobile and embeddable database management systems should have is set out: be distributed with the application; minimize main and secondary memory usage; allow only the necessary DBMS components to be included; support main memory storage; be portable; run on mobile devices; and synchronize data with backend DBMSs.

In [12] a survey of different storage options in mobile devices is carried out, being these: HTML5, from the use of the WebKit framework, enabling the use of the *localStorage* API; SQLite, a library that encapsulates SQL functionality and stores the information in a local file; cloud storage, using services such as *Apple iCloud, Google Drive, Dropbox, Amazon S3*; device-specific storage, such as *Shared Preferences* on Android and *Core Data* on iOS, in addition to the previous options, present in both systems.

In [13] the architecture of the Android platform is described, and the architecture and the way in which applications are executed, among other issues, is discussed. Finally, an introduction to the use of SQLite on Android is given.

In [14] the advantages and disadvantages of using cloud computing in an integral way with mobile applications is discussed, mentioning storage, backups, and data redundancy as advantages, among others.

In the previously described works, although issues related to data storage in mobile devices are analysed, there is no analysis of the different database management systems for mobile devices that would allow the software engineer to select the most appropriate DBMS to use to solve a given problem, an aspect intended to cover with the presentation of this article.

3   Databases for Mobile Devices

A survey of relational DBMSs (RDBMSs) and NoSQL DBMSs that can be used embedded in mobile applications is presented. The search for DBMSs to be analysed was carried out through the search engines Google and Google Scholar, using the terms "mobile database", "mobile dbms", among others. A Google search was also performed using the term "mobile site:db-engines.com/en/system", since the DB-Engines site uses that *path* for pages describing features of each DBMS surveyed by the site.

DB-Engines is a project to collect and present information about DBMSs, created and maintained by solidIT, an Austrian company specialized in software development, consulting and training for data-centric applications. The site prepares a monthly ranking based on the score obtained by each DBMS surveyed according to its popularity, defined by different parameters [3].

3.1   Relational DBMSs

The existing relational DBMSs or RDBMSs for mobile devices, sorted according to the ranking prepared by DB-Engines [3] are:

1. SQLite
2. Interbase
3. SAP SQL Anywhere
4. SQLBase

SQLite is a library that implements a self-contained (embedded) database engine. It is Public Domain licensed, and can be used both in native application development, either on Android or iOS, as well as cross-platform application development in hybrid, interpreted or cross-compile approaches [15].

Interbase is an embedded, commercially licensed, SQL-compliant RDBMS. As for its use in mobile application development, it can be used in native development, both in Android and iOS [16].

SAP SQL Anywhere integrates a suite of relational DBMSs and synchronization technologies for servers, desktop and mobile environments [17]. It is available for use in developing native mobile applications on both Android and iOS.

SQLBase is an RDBMS developed by the company Opentext. It is licensed for commercial use. It is available for the development of native mobile applications, on Android and iOS [18].

3.2 NoSQL DBMSs

The term NoSQL is used to refer to databases of distinct models, different from the relational model. Among the different existing NoSQL database models, the document store model represents the most widely used today [3]. Of the existing NoSQL DBMSs for mobile devices, most correspond to the document store model.

The existing NoSQL DBMSs for mobile devices, sorted according to the DB-Engines ranking are:

1. Couchbase Lite (Document Store)
2. Firebase Realtime Database (Document Store)
3. Realm (Document Store)
4. Google Cloud Firestore (Document Store)
5. Oracle Berkeley DB (Key-Value Store)
6. PouchDB (Document Store)
7. LiteDB (Document Store)
8. ObjectBox (Object-Oriented Store)
9. Sparksee (Graph Store)

Document Store DBMSs
Couchbase Lite is an embedded DBMS, which uses JSON as the document format. Being part of the Couchbase package, it is possible to use the *Sync Gateway* system to synchronize data with remote [19] databases. It is dual-licensed, and can be used for Android and iOS native application development, and cross-platform development, in hybrid, interpreted, and cross-compiled approaches [3].

Firebase Realtime Database is a cloud-based DBMS, which stores data in a single JSON, and features real-time data synchronization with all connected clients, and maintaining data available even offline. It is available for use in native application development in both Android and iOS, and cross-platform mobile application development in hybrid, interpreted, and cross-compiled approaches [20].

Realm is an embedded DBMS that uses an object-oriented data model. It can be used standalone, or synchronized with a MongoDB database. It is available for native Android and iOS application development, and cross-platform mobile application development in interpreted and cross-compiled approaches [3,21].

Google Cloud Firestore is a cloud-based DBMS, which stores data in JSON documents and features real-time data synchronization with clients, keeping data available even offline. It can be used in native applications development on Android and iOS, and in the development of cross-platform mobile applications with hybrid, interpreted, and cross-compiled approaches [22].

PouchDB is a javascript library that implements a DBMS inspired by CouchDB, and uses its synchronization protocol. It is distributed under Apache 2.0 license and is available for cross-platform mobile application development under the hybrid and interpreted approaches [23].

LiteDB is a library distributed as a single DLL file under MIT license. It uses BSON documents to store information. It is available for cross-platform mobile application development in Xamarin (cross-compiled) [24].

DBMSs of Other Types

Oracle Berkeley DB is a family of embedded key-value store databases. It is open source licensed and is available for Android and iOS native application development [3,25].

ObjectBox is an object-oriented DBMS for mobile devices and IoT. It is licensed under Apache 2.0, and is available for native mobile app development on Android and iOS, and cross-platform on Flutter. It has a synchronization and cloud storage service [3,26].

Sparksee is a graph DBMS. It is distributed under commercial license and has free licenses for educational and research porpoises. It can be used in native application development on both Android and iOS [3,27].

4   Experimentation

To carry out the experimentation, DBMSs are selected under the following conditions, defined for this work: being able to be used without a commercial license; beign able to perform synchronization with remote databases; being able to be used in mobile applications developed natively on the main platforms (Android and iOS); being able to be used in mobile applications developed with different cross-platform approaches in known development frameworks (React Native, NativeScript, Ionic Framework, Xamarin, Flutter); being in the first three positions according to the ranking by DBMS model elaborated by DB-Engines.

Based on the characteristics of each DBMS, and the criteria previously stated, SQLite, Couchbase Lite, Firebase Realtime Database and Realm were selected for the comparative experimental analysis.

To carry out the analysis, three mobile applications were partially developed on the Android platform, each using the selected DBMSs. The application consists of a contact book that meets the following requirements:

Functional Requirements:

R1 The application must allow to create a new contact, with the following data: Last name (required); First name (required); Date of birth (optional); Emails (zero or more); Phones (zero or more; for each Phone is stored: Number —required— and Type —required, one of the following options must be selected: Mobile, Home, Work, Other—)

R2 The application must allow to list the stored contacts sorted by last name and first name in ascending order.

R3 The application must allow to filter the stored contacts, using a single search term given, listing the contacts that partially or totally match the search term in some of its fields. The filtered contacts will be displayed sorted by Last Name and First Name in ascending order.

Non-functional Requirements:

R4 All information must be stored locally in the database.

For the development of the application, a single diagram corresponding to the database conceptual model is defined, using the Entity-Relationship Model, which is then derived to the corresponding logical and/or physical models according to each database model and DBMS.

The analysis will be performed from the point of view of the software engineer's experience in the development of each application, considering the complexity of implementation for the use of the corresponding DBMS, in terms of factors such as installation/configuration, implementation of required components—e.g., data structures, classes—, definition/execution of queries, among others.

Given the objective of the analysis, it is not considered relevant that the defined data schema is better adapted to a specific database model, since the interest of this work is focused on analyzing the implementation of the different specific characteristics of each database model.

The requirements and data schemas defined are intended to implement the specific features of the selected database models: *relations* in the relational model; *embedded documents, arrays of scalar types,* and *arrays of documents* in the document store NoSQL model.

4.1 SQLite

For the experimentation using SQLite as DBMS, the diagram corresponding to the conceptual model is derived to the physical model (Fig. 1).

For the use of SQLite in the development of mobile applications, Android provides two ways to manage databases in said DBMS. The first one (recommended by the official documentation) is using Room, a persistence library that works as a SQLite abstraction layer; the second one is using the SQLite API directly [28]. The chosen way for experimentation is the first one.

Installation/Configuration. Room is installed by adding the necessary dependencies in the *gradle* file.

Contacto(<u>id</u>, apellido, nombre, fecha_nacimiento?, empresa?, calle?, nro?, piso?, depto?)
Telefono(<u>id</u>, numero, tipo, contacto_id *(FK)*)
Email(<u>id</u>, email, contacto_id *(FK)*)

Fig. 1. Diagram corresponding to the physical model for SQLite

Data Schema Definition. Room is used by defining classes and object interfaces to which certain annotations must be defined.

The library has three main components: *entities*, which are classes that represent the entities of the model, and which represent the tables of the relational model. These classes should be annotated with @Entity.

The second main component of Room are the *DAOs*, defined from interfaces with the annotation @Dao, which are used to perform queries on the table that represents each entity, allowing to create, retrieve, update, and delete entities.

The third component is the *database*, which serves as the primary access point for the underlying connection to the relational database. An abstract class must be defined that extends from the RoomDatabase class, and annotated with @Database.

In the relationships among the different entities, the foreign key restrictions must be noted, indicating for each one the entity and property o properties it refers to, and the property on which it applies.

Due to the impossibility of referencing objects in Room, properties whose type does not correspond to a scalar data type (numeric, strings, and boolean types) must be converted in order to be stored in the database. For this purpose, two conversion methods must be defined for each non-scalar data type to be persisted.

Data Insertion. To satisfy the functional requirement *R1*, the tuples corresponding to the contact to be created, and the contact's telephone numbers and emails must be inserted. To do so, methods must be defined in the interfaces corresponding to the *DAOs* annotated with @Insert.

Data Retrieval. The application developed as part of the experiment should list all stored contacts (*R2*), and contacts that partially or completely match a search term (*R3*). For tuple retrieval queries, methods must be defined in the interface corresponding to the *DAO* annotated with @Query, whose annotation value is the SQL query. To filter from a search term it is necessary to parameterize it, which is achieved by defining a parameter in the method, whose value can be referenced in the query by prefixing the parameter name with a colon.

In cases where it is necessary to retrieve information from multiple interrelated tables, a class with properties whose type corresponds to the entities involved must be implemented.

Source Code. It is published in [29].

4.2 Couchbase Lite

A particular feature of document store databases is that they offer the possibility of storing unstructured or semi-structured data. It is also possible to partially or completely define a schema for the documents.

For the implementation, no schema is defined for the database, but the documents are structured according to the physical model in Fig. 2.

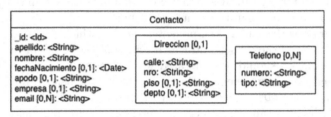

Fig. 2. Diagram corresponding to the physical model for Couchbase Lite

Installation/Configuration. Couchbase Lite is installed by adding the necessary dependencies in the *gradle* file.

Data Schema Definition. Data management is performed using classes already defined by Couchbase Lite, including `MutableDocument`, `MutableArray`, `MutableDictionary`, among others. Objects of these classes are used to define the structure of each document to be stored in the database; each document can have a unique structure.

While it is possible to define a class structure for an object-oriented implementation, the hydration of the objects must also be implemented.

Data Insertion. To add documents, instances of the class `MutableDocument` must be created and the document fields must be defined (optional fields that do not have a value are not defined).

To embed documents, the class `MutableDictionary` is used, whether they correspond to a single document or to an array of documents.

In the case of arrays, either embedded documents or scalar data types, the class `MutableArray` must be used to define them. Therefore, an array of embedded documents is defined as a `MutableArray` whose values are instances of `MutableDictionary`.

Data Retrieval. To perform data retrieval queries, an object of the class `QueryBuilder` must be used, to which messages must be sent to define expressions to establish conditions on document properties, sorting criteria, among others. The data types returned by the query execution are immutable variants of those used for inserts (`Dictionary`, `Array`; documents are encapsulated in objects of class `Result`).

Source Code. It is published in [30].

4.3   Firebase Realtime Database

The implementation of Firebase Realtime Database as a DBMS uses the same physical model defined for Couchbase Lite, expressed in Fig. 2. The official DBMS documentation states that since the information is stored as a single JSON tree, it is convenient to avoid nesting data. For this reason, the physical model used for this work would not be optimal for a solution, but it is considered adequate to perform the intended analysis.

Installation/Configuration. The installation of the Firebase Realtime Database SDK is done by adding the corresponding dependencies in the *gradle* file.

To configure the connection to the cloud database, the *google-services.json* configuration file must be generated through the Firebase console and added to the project's */app* directory.

Data Schema Definition. Data management in Firebase Realtime Database is performed from reads or writes in which a URI is specified that allows retrieving or modifying a data sub-tree, called *reference*. For example, in the physical model defined, to retrieve all the contacts stored, a read operation must be made to the reference */contacts*, and to retrieve the contact with identifier 3, to */contacts/3*.

Write operations can be performed by defining values for a specific reference. The values must correspond to one of the following *Java* data types: `String`, `Long`, `Double`, `Map<String, Object>`, `List<Object>`, `Object`.

In the case of the `Object` type, objects of any class can be used as long as it has a public constructor method that does not receive any arguments, and public *get* methods to access its properties.

With this in mind, a set of classes representing the data model was defined, with the only special consideration being the generation of an identifier for each new contact (using the *UUID* class defined in the *java.util* package).

Data Insertion. Data insertion in Firebase Realtime Database is done by defining a value for a given reference. In the case of write operations defining an object as a value, the object is converted to JSON, taking as keys' names the class properties' names. Only properties with non-null values are written.

If the device is not connected to the Internet when performing write operations, it is possible that the write operations will be stored on the device until connectivity is restored and synchronization with the cloud database can be performed.

Data Retrieval. Analogous to data insertion, data retrieval requires reading a certain reference. Each read operation provides a reference's *snapshot* (an object), from which data can be obtained. In the case of data collections, the retrieved snapshot contains a collection of snapshots, each corresponding to a piece of data. In the case of objects, an instance of a given class can be obtained by supplying the name of the class to the snapshot.

To satisfy the functional requirement $R3$, the filtering and sorting of the data after retrieving it from the database must be implemented, since the DBMS does not allow partial searches on the stored data, nor sorting by more than one criterion.

If the device loses connectivity, changes made after the loss of connectivity are persisted locally on the device, so data retrieval shows changes made locally.

The non-functional requirement $R4$ is considered to be accomplished, since the database is considered to be used by only one user, and it can be configured to also store locally a given reference. So in the case of connectivity loss, all data remains stored locally in the device, even if its not synchronized with the cloud (which will be eventually, when connectivity is restored).

Source Code. It is published in [31].

4.4   Realm

For the implementation of the application using Realm, the same physical model defined for the previous document store DBMSs, presented in Fig. 2, is used, since, although there are differences between the DBMSs, the structures defined in the model also apply to Realm.

Installation/Configuration. Realm is installed by adding the necessary dependencies in the *gradle* file.

Data Schema Definition. Realm is used by defining classes that represent the documents of the model, which must be defined as subclasses of `RealmObject` or that implement the `RealmModel` interface, whether they are embedded documents or not.

For *top-level* documents, a property with type `ObjectId` must be defined and annotated with `@PrimaryKey`. Classes corresponding to documents that are used embedded in other documents, either directly or as an array of documents, must be annotated with `@RealmClass(embedded = true)`.

In cases where documents are defined with arrays of scalar data type values or arrays of documents, the properties' types must be `RealmList<T>`, where `T` is the scalar data type or the class corresponding to the embedded document.

Data Insertion. The write operations in the database in Realm are performed through transactions, which are defined by invoking the `executeTransaction` method of the database instance, where a lambda expression is passed as a parameter in which the behavior is defined. In the body of the expression, an instance of the class corresponding to the document to be inserted must be obtained and the values of the properties to be persisted must be defined.

Data Retrieval. To perform queries to retrieve stored documents, an instance of `RealmQuery<T>` must be obtained, where `T` corresponds to the document class to be retrieved, invoking the `where` method of the database instance, passing as an argument the name of the `T` class.

The `RealmQuery` class allows to invoke different methods for filtering and sorting, among others, and to obtain the documents encapsulated in an instance of `RealmResults<T>`.

Source Code. It is published in [32].

5   Results Analysis

5.1   SQLite

In the experimentation using SQLite as DBMS to persist the information, it was decided to use the Room library, which represents an abstraction layer over SQLite. In the implementation performed, the following advantages stand out: (1) the class structure that needs to be defined leads to working in an organized and systematic way; (2) the definition of the DB schema from the classes that make up the entities of the model, so it is not necessary to define it with SQL statements; (3) the flexibility provided in defining queries to retrieve information, since only the query must be defined as the value of an annotation, including parameterized queries; (4) the ease of defining type conversions that cannot be stored directly in the database; (5) the ease with which tuples are inserted.

In addition, some disadvantages of using Room are: (1) the need to use auxiliary structures to obtain tuples resulting from products (*joins*) between tables; (2) the need to define a converter for a widely used data type such as the *Date* data type.

5.2   Couchbase Lite

In Couchbase Lite, considering the experimentation performed, the following advantages could be observed: (1) flexibility in the definition of the document structure, as it can be defined completely dynamically; (2) the use of structures defined by Couchbase Lite for persistence and information retrieval.

Can be considered as disadvantages: (1) it is not possible to group documents into collections; (2) the implementation of complex queries requires a large amount of code to be defined, which difficult its comprehension; (3) the implementation of queries that reference values in arrays is complex due to the need to define variables that reference each value in the array; (4) the need to implement object hydration in case of developing the application using the object-oriented programming paradigm;

5.3   Firebase Realtime Database

In the implementation with Firebase Realtime Database it is possible to list the following advantages: (1) the definition of document structure from the definition of object classes; (2) flexibility in the definition of the document structure; (3) the ease with which data can be inserted and retrieved.

In turn, the following are considered disadvantages: (1) the impossibility of filtering data based on partial matching of values from a search term, and (2) the impossibility of sorting data on the basis of two or more criteria.

5.4   Realm

The Realm implementation offers the following advantages: (1) the definition of the document structure through the definition of classes; (2) the possibility of grouping documents into collections; (2) the implementation of queries is concise and clear, especially since it is not necessary to define variables to reference fields in embedded documents or in arrays of embedded documents.

Also, the impossibility of referencing values corresponding to arrays of scalar types in queries was found to be a disadvantage.

5.5   Analysed DBMSs: Comparative Table

Table 1 summarizes the aspects evaluated in SQLite, Couchbase Lite, Firebase Realtime Database, and Realm. Each aspect was rated on a five-value scale: *very low, low, medium, high, very high*. The aspects evaluated are:

- Complexity: expresses the level of difficulty to perform the task;
- Flexibility: applies only to the definition of the data schema. Refers to how flexible the schema is to adapt to changes;
- Code readability: expresses the level of difficulty to understand the source code;
- Integration with OOP (Object-Oriented Programming): level of interaction of the DBMS data types with the objects defined in the application;
- Support of non-scalar data types: level of difficulty in the persistence of non-scalar data types, i.e. data types other than numeric, strings, and boolean types..

The best possible rating is *very high* for all the aspects analysed, except for Complexity, whose best rating is *very low*.

Table 1. Aspects evaluated in the DBMSs

Tasks	Aspects	SQLite	Couchbase lite	Firebase realtime database	Realm
Installation/ Configuration	Complexity	Very low	Very low	Low	Very low
Data schema definition	Complexity	Low	Very low	Very low	Very low
	Flexibility	Low	Very high	Very high	Medium
	Code readability	High	Very low	Very high	Very high
	Integration with OOP	High	Very low	Very high	High
	Support of non-scalar data types	Very low	Low	Very high	Low
Data insertion	Complexity	Low	Medium	Very low	Low
	Code readability	Very high	High	Very high	High
	Integration with OOP	Very high	Low	Very high	High
Data retrieval	Complexity	Low	Very high	Medium	Very low
	Code readability	Very high	Very low	Medium	Very high
	Integration with OOP	High	Very low	Very high	Very high

6   Conclusions and Future Work

This paper attempts to address the problem of choosing a suitable DBMS that can be embedded in a mobile application, according to the problem to be solved.

The exponential increase in the volume of managed data, including structured, semi-structured, and unstructured data, led to the emergence of new database models, called NoSQL databases, to facilitate the massive storage of semi-structured and unstructured data.

On the other hand, improvements in hardware performance of mobile devices are leading to more and more information being managed by these devices, and to the emergence of new database management systems that are installed on these devices.

As a result, this work analyzes different aspects related to the installation and/or configuration, implementation of required components, definition and/or execution of modification queries and data retrieval, to assist the software engineer in the selection of a suitable DBMS to be embedded in mobile applications according to the problem to be solved.

From the experimentation and analysis carried out, it can be considered that the problem to be solved is a determining factor in the choice of a DBMS. SQLite, Firebase Realtime Database and Realm would be the most suitable because of their low complexity, high code readability and high integration with object-oriented programming paradigm structures, in the tasks corresponding to data schema definition, data insertion and data retrieval, although in this last aspect, Firebase Realtime Database may not be suitable if the problem to be solved requires partial searches on stored values, or sorting the results by more than one criterion. SQLite and Realm have a low flexibility in the definition of the data schema, where Couchbase Lite and Firebase Realtime Database would be more suitable.

In situations where you need to perform partial searches on the managed data, or you need to sort results by more than one criterion, Firebase Realtime Database would not be suitable. In situations where great flexibility in the data schema is not required, SQLite and Realm would be the most suitable. On the other hand, in situations where great flexibility in the data schema is required, Couchbase Lite would represent a better option. However, considering only the flexibility in the data schema, Firebase Realtime Database is suitable regardless of the required flexibility, since it easily adapts to the needs of the problem to be solved. In the proposed experimentation, which requires performing partial match searches on the managed data, sorting data by more than one criterion, and does not require a flexible data schema, SQLite and Realm are more suitable, the former being considered the best option, since the problem posed is better suited to the relational model.

Finally, several lines of research are proposed as possible future work:

1. extending the experimentation by adding the non-functional requirement of synchronizing the mobile application database with a backend database;

2. analyzing the impact of using SQLite, Couchbase Lite, Firebase Realtime Database or Realm on non-functional requirements that are critical to the success of the application, such as main and secondary memory usage, query execution performance, and power consumption;
3. extending the analysis using other DBMSs, being particularly of interest those DBMSs that are available for use in both Android and iOS, and in known frameworks for cross-platform mobile applications development.

References

1. Kristi L Berg, Tom Seymour, and Richa Goel. History of databases. International Journal of Management & Information Systems (IJMIS), 17(1):29–36, 2013
2. Burton Grad and Thomas J Bergin. Guest editors' introduction: History of database management systems. IEEE Annals of the History of Computing, 31(4), 3–5, 2009
3. Db-engines ranking - populariry ranking of database management systems. https://db-engines.com/en/ranking. Accedido por última vez: 17/02/2021
4. Martin Campbell-Kelly and Daniel D Garcia-Swartz. From mainframes to smartphones: a history of the international computer industry, volume 1. Harvard University Press, 2015
5. List of countries by smartphone penetration - wikipedia. https://en.wikipedia.org/wiki/List_of_countries_by_smartphone_penetration#2013_rankings. Accedido por última vez: 17/02/2021
6. Cell phone sales worldwide 2007–2020 | statista. https://www.statista.com/statistics/263437/global-smartphone-sales-to-end-users-since-2007/. Accedido por última vez: 17/02/2021
7. Lisandro Delia, Nicolas Galdamez, Pablo Thomas, Leonardo Corbalan, and Patricia Pesado. Multi-platform mobile application development analysis. In 2015 IEEE 9th International Conference on Research Challenges in Information Science (RCIS), pages 181–186. IEEE, 2015
8. Spyros Xanthopoulos and Stelios Xinogalos. A comparative analysis of cross-platform development approaches for mobile applications. In Proceedings of the 6th Balkan Conference in Informatics, pages 213–220, 2013
9. Luciano Marrero, Verena Olsowy, Pablo Javier Thomas, Lisandro Nahuel Delía, Fernando Tesone, Juan Fernández Sosa, and Patricia Mabel Pesado. Un estudio comparativo de bases de datos relacionales y bases de datos nosql. In XXV Congreso Argentino de Ciencias de la Computación (CACIC)(Universidad Nacional de Río Cuarto, Córdoba, 14 al 18 de octubre de 2019), 2019
10. Fernando Tesone, Pablo Javier Thomas, Luciano Marrero, Verena Olsowy, and Patricia Mabel Pesado. Un análisis experimental de sistemas de gestión de bases de datos para dispositivos móviles. In XXVII Congreso Argentino de Ciencias de la Computación (CACIC)(Modalidad virtual, 4 al 8 de octubre de 2021), 2021. ISBN: 978-987-633-574-4
11. Anil Nori. Mobile and embedded databases. In Proceedings of the 2007 ACM SIGMOD International Conference on Management of data, pages 1175–1177, 2007
12. Qusay H Mahmoud, Shaun Zanin, and Thanh Ngo. Integrating mobile storage into database systems courses. In Proceedings of the 13th annual conference on Information technology education, pages 165–170, 2012

13. Sunguk Lee. Creating and using databases for android applications. International Journal of Database Theory and Application, 5(2), 2012
14. Ahmed Alzahrani, Nasser Alalwan, and Mohamed Sarrab. Mobile cloud computing: advantage, disadvantage and open challenge. In Proceedings of the 7th Euro American Conference on Telematics and Information Systems, pages 1–4, 2014
15. About sqlite. https://www.sqlite.org/about.html. Accedido por última vez: 21/02/2021
16. Interbase - embarcadero website. https://www.embarcadero.com/es/products/interbase. Accedido por última vez: 03/03/2021
17. Sap sql anywhere | rdbms for iot & data-intensive apps | technical information. https://www.sap.com/products/sql-anywhere/technical-information.html. Accedido por última vez: 03/03/2021
18. Opentext gupta sqlbase. https://www.opentext.com/products-and-solutions/products/specialty-technologies/opentext-gupta-development-tools-databases/opentext-gupta-sqlbase. Accedido por última vez: 03/03/2021
19. Lite | couchbase. https://www.couchbase.com/products/lite. Accedido por última vez: 21/02/2021
20. Firebase realtime database | firebase realtime database. https://firebase.google.com/docs/database. Accedido por última vez: 21/02/2021
21. Home | realm.io. https://realm.io/. Accedido por última vez: 22/02/2021
22. Cloud firestore | firebase. https://firebase.google.com/docs/firestore/. Accedido por última vez: 22/02/2021
23. Pouchdb, the javascript database that syncs! https://pouchdb.com/. Accedido por última vez: 21/02/2021
24. Litedb: A.net embedded nosql database. http://www.litedb.org/. Accedido por última vez: 03/03/2021
25. Oracle berkeley db. https://www.oracle.com/database/technologies/related/berkeleydb.html. Accedido por última vez: 03/03/2021
26. Mobile database | android database | ios database | flutter database. https://objectbox.io/mobile-database/. Accedido por última vez: 03/03/2021
27. Sparsity-technologies: Sparksee high-performance graph database. http://sparsity-technologies.com/#sparksee. Accedido por última vez: 03/03/2021
28. Descripción general del almacenamiento de archivos y datos. https://developer.android.com/training/data-storage. Accedido por última vez: 21/02/2021
29. Código fuente de experimentación realizada en android con sqlite y room. https://github.com/ftesone/tesina-room/tree/springer-2022. Accedido por última vez: 03/03/2022
30. Código fuente de experimentación realizada en android con couchbase lite. https://github.com/ftesone/tesina-couchbase/tree/springer-2022. Accedido por última vez: 03/03/2022
31. Código fuente de experimentación realizada en android con firebase realtime database. https://github.com/ftesone/android-firebase/tree/springer-2022. Accedido por última vez: 03/03/2022
32. Código fuente de experimentación realizada en android con realm. https://github.com/ftesone/android-realm/tree/springer-2022. Accedido por última vez: 03/03/2022

Hardware Architectures, Networks, and Operating Systems

Service Proxy for a Distributed Virtualization System

Pablo Pessolani$^{(\boxtimes)}$ ⓘ, Marcelo Taborda, and Franco Perino

Department of Information Systems Engineering, Universidad Tecnológica Nacional – Facultad Regional Santa Fe, Santa Fe, Argentina
{ppessolani,mtaborda,fperino}@frsf.utn.edu.ar

Abstract. Most cloud applications are composed by a set of components (microservices) that may be located in different virtual and/or physical computers. To achieve the desired level of performance, availability, scalability, and robustness in this kind of systems is necessary to maintain a complex set of infrastructure configurations.

Another approach would be to use a Distributed Virtualization System (DVS) that provides a transparent mechanism that each component could use to communicate with others, regardless of their location and thus, avoiding the potential problems and complexity added by their distributed execution. This communication mechanism already has useful features for developing distributed applications with replication support for high availability and performance requirements.

When a cluster of backend servers runs the same set of services for a lot of frontend clients, it needs to present a single entry-point for them. In general, an application proxy is used to meet this requirement with auto-scaling and Load Balancing features added. Auto Scaling is the mechanism that dynamically monitors the load of the cluster nodes and creates new server instances when the load is greater than the threshold of maximum CPU usage, or it removes server instances when the load is less than the threshold of minimum CPU usage. Load balancing is another related mechanism that distributes the load among server instances to avoid that some instances are saturated and others unloaded. Both mechanisms help to provide better performance and availability of critical services.

This article describes the design, implementation, and evaluation of a service proxy with Autos Scaling and Load Balancing features in a DVS.

Keywords: Auto scaling · Load balancing · Distributed systems

1 Introduction

Nowadays, applications developed for the cloud demand more and more resources, which cannot be provided by a single computer. To increase their computing and storage power, as well as to provide high availability and robustness they run in a distributed environment. Using a distributed system, the computing and storage capabilities could be extended to several different physical machines (nodes). Although there are various distributed processing technologies, those that offer simpler ways of implementation, operation, and maintenance are highly valued. Also, technologies that provide a Single System Image (SSI) are really useful because they abstract the users and programmers from issues such as the location of processes, the use of internal IP addresses, TCP/UDP ports, etc., and more importantly, because they hide failures by using replication mechanisms. A Distributed Virtualization System (DVS) is an SSI technology that has all of these features [1]. A DVS offers a distributed virtual runtime environments in which multiple isolated applications can be executed. The resources available to the DVS are scattered in several nodes of a cluster, but it offers aggregation capabilities (allows multiple nodes of a cluster to be used by the same application), and partitioning (allows multiple components of different applications to be executed in the same node) simultaneously. Each distributed application runs within an isolated domain or execution context called a Distributed Container (DC). Figure 1 shows an example of a topological diagram of a DVS cluster.

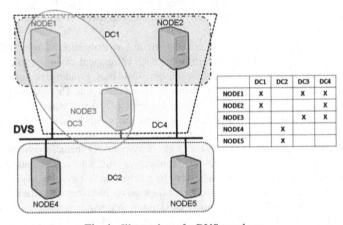

	DC1	DC2	DC3	DC4
NODE1	X		X	X
NODE2	X			X
NODE3			X	X
NODE4		X		
NODE5		X		

Fig. 1. Illustration of a DVS topology.

A problem that must be considered by a distributed application refers to the location of a certain service used by an external or internal client, or by another component of the distributed application itself. One way to solve this problem would be to use existing Internet protocols. With the DNS protocol, the IP address of the server can be located in the IP network, and with ARP the MAC address of the server can be located within a LAN. However, one issue that must be considered, when working with a cluster, is that the network and its nodes may fail, preventing the continuity in the delivery of a given service.

Current DevOps practices make developers responsible for including with their software the description and configuration of infrastructure necessary to support it. This includes computing configuration (as computing nodes), storage, networking and security. They must consider routing and firewall rules, replication and backup strategies, Auto Scaling and Load Balancing specifications, etc.

This practice diverts them from their main goal, which is to improve their own applications, forcing them to learn about technologies at other levels of abstraction and for which deep knowledge is required. While there are several tools for deploying Infrastructure as Code (IaaC) that complement others used for Continuous Integration and Continuous Deployment (CI/CD), the specifications in the configuration files (markup languages are generally used) must be mastered.

In a DVS, the infrastructure level is clearly separated from the application level, as well as the responsibilities of those who manage them.

When a cluster of backend servers runs the same set of applications for a lot of frontend clients, it needs to present a single entry-point for their services. In general, an Application Proxy (AP), also known as Reverse Proxy, is used to meet this requirement with auto-scaling and Load Balancing features added. Auto Scaling is the mechanism that dynamically monitors on the load on the back-end servers and creates new server instances when the load is higher than a threshold of maximum resource usage (generally CPU usage), or it removes server instances when the load is lower than another threshold. Load Balancing is another related mechanism that distributes the load among backend server instances to avoid that some instances are saturated and others unloaded. Both mechanisms help to provide better performance and availability of critical services. This technology is widely used in certain scenarios such as web applications, where end-users send requests from their devices as clients, and the SP is the component that establishes sessions with backend servers, thus distributing, balancing, and orchestrates services between internal services and microservices [2]. To distinguish between these two usages scenarios, in this article Application Proxy (AP) or Application Gateway refers to the former, and Service Proxy (SP) or Internal Gateway refers to the latter (Fig. 2).

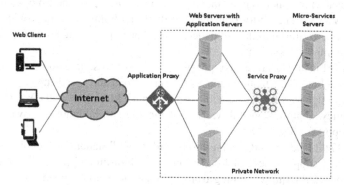

Fig. 2. Example of a MicroService architecture (MSA)

This article is in continuation of a previous work presented in [3] with significant improvements such as more detailed design explanation, functional and performance testing and extended discussion. It presents the design, implementation, and experimental proves the capabilities of an SP with Auto Scaling, and Load Balancing features for a DVS. Therefore, the project focused on building an operational prototype of an SP for the DVS (not a commercial-class one), relegating performance enhancements and high availability features for future works.

The rest of the article is organized as follows: Sect. 2 refers to related works. Section 3 provides an overview of background technologies and Sect. 4 describes the design and implementation of the SP for the DVS (referred to as DVS-SP). Section 5 presents the tests for the SP performance evaluation and finally, the conclusions and future works are summarized in Sect. 6.

2 Related Works

APs and SPs are not unknown topics by the scientific community, so a lot of research and development works have previously been carried out, but most refer to IP environments. Although there are distributed approaches, DVS-SP is a centralized one therefore such as several of the most used in Cloud environments. However, it was the first step to a distributed Service Proxy of a work in progress.

A very popular HTTP server and reverse proxy is NGINX [4]. It is free, open-source, and well known for its high performance, stability, rich feature set, and low resource consumption. NGINX can handle tens of thousands of concurrent connections and provides caching when using the *ngx_http_proxy_module* module and supports Load Balancing and fault tolerance. The *ngx_http_upstream_module* module allows for the *nginx groups* of backend servers to distribute the requests coming from clients.

HAproxy [5] (stands for High Availability Proxy) is an HTTP reverse-proxy. It is a free, open-source, reliable, high-performance load balancer and proxy software for TCP and HTTP-based applications. It is also an SSL/TLS tunnel terminator, initiator, and off-loader, and provides HTTP compression and protection against DDoS. It can handle tens of thousands of concurrent connections by its event-driven, non-blocking engine.

HAproxy was designed for high availability, Load Balancing and provides redirection, server protection, logging, statistics, and other important features for large-scale distributed computing systems.

Varnish [6] is an HTTP reverse proxy software (open-source and free) which uses server-side caching. This feature reduces the server load and speeds up the web content delivery. Although Varnish is extensible using Varnish Modules (VMODs) and provides caching, policy, analytics, visibility, Load Balancing and mitigation for web traffic, it lacks of support for SSL/TLS.

Traefik [7] it is an HTTP reverse-proxy with Load Balancing for deploying microservices which supports several Load Balancing algorithms. Traefik is used to route HTTP(S) and TCP requests coming from Internet clients to frontend servers. Its main features are: a) service discovery which automatically manages its configuration; b) providing access logs and metrics; and c) supporting a wide range of protocol such as WebSockets, HTTP/2, HTTPS/SSL and GRPC.

Apache Traffic Server (ATS) [8] is a caching forward and reverse-proxy server, which features such as load balancing, keep-alive, logging, filtering, or anonymizing of content requests. It is also extensible via APIs through custom plugins. Although ATS has demonstrated the higher performance (higher throughput and lower response time) than Nginx and Varnish [9], it has had several security vulnerabilities [10].

3 Background Technologies

This section presents the products and tools that have been studied and analyzed as technological support for the design and implementation of the DVS-SP prototype.

3.1 M3-IPC

The DVS provides programmers with an advanced IPC mechanism named M3-IPC [11] in its Distributed Virtualization Kernel (DVK) which is available at all nodes of the DVS cluster. M3-IPC provides tools to carry out transparent communication between processes located at the same (local) node or in other (remote) nodes. To send messages and data between processes of different nodes, M3-IPC uses Communications Proxies (CPs) processes. CPs act as communication pipes between pairs of nodes.

M3-IPC processes are identified by *endpoints* that are not related to the location of each process, and then it does not change after a process migration. This feature becomes an important property that facilitates application programming, deployment, and operation. An *endpoint* can be allocated by a process or by a thread and must be unique in each DC.

M3-IPC supports message transfers (which have a fixed size) and blocks of data between endpoints. If the sender and receiver endpoints are located on the same node, the kernel copies the messages/data between the processes/threads which own the endpoints. If the sender and receiver are located in different nodes, CPs are used to transfer messages and data between nodes, and the DVK copies those messages/data between the CPs and the processes/threads.

3.2 Group Communication System (GCS)

To exchange information between a group of processes that run on several nodes or in the cloud, communication mechanisms with characteristics such as reliability, fault tolerance, and high performance are required. Several tools offer these features such as Zookeeper [12], Raft [13] or the Spread Toolkit [14].

The Spread Toolkit was chosen for the DVS-SP development because it is a well-known GCS used by the authors' research group in other projects. On Spread Toolkit, two kinds of messages are distinguished. Regular messages: sent by a group member, and; Membership messages: sent by the Spread agent running on each node.

Regular messages can be sent by members using the provided APIs for broadcast (multicast) them to a group. However, unicast messages could be sent to a particular member. Membership messages are sent by Spread to notify members about a membership change, such as the joint of a new member, the disconnection of a member, or a network change. Network changes can be the crash of a node (or a set of nodes), a network partition, or a network merging after a partition.

The Spread toolkit provides reliable delivery of messages (even in the event of network or group member failures) and the detection of failures of members, or the network. It also supports different types of ordering in message delivery, such as FIFO, Causal, Atomic, etc. making it an extremely flexible tool for the development of reliable distributed systems.

The Spread Toolkit is based on the group membership model called Extended Virtual Synchrony (EVS) [15], tolerating network partition failures and network merging after a partition, node failures, process failures and process restarts.

4 Design and Implementation of the DVS-SP

The design of the DVS-SP started proposing its architecture, describing its components and the relations between them. The active components are (Fig. 3):

- The Main Service Proxy (MSP) reads a configuration file, initializes all data structures, and starts the other component threads.
- For each Frontend Client Node (specified in the configuration file), a pair threads is started for the CPs. The Client Sender Proxy (CSP) thread sends messages from the DVS-SP to the Client node. The Client Receiver Proxy (CRP) thread receives messages from the Client node.
- Similarly, for each Backend Server Node, a pair threads is started for the CPs. The Server Sender Proxy (SSP) thread sends messages from the DVS-SP to the Server node. The Server Receiver Proxy (SRP) thread receives messages from the Server node.
- A Load Balancer Monitor (LBM) thread receives notifications about changes in the load state from the Load Balancer Agents (LBA) running on each backend server node.
- Each Client and Server node uses the Node Sender Proxy (NSP) and the Node Receiver Proxy (NRP) to communicate with the DVS-SP proxies.

Proxy messages differ from application messages. Proxy messages are the transport of single application messages (like a tunnel). Each proxy message can pack a single or a batch of application messages, a single block of raw data, acknowledgment messages, or a proxy HELLO message. A HELLO proxy message is sent out periodically between proxies to confirm the network reachability and the remote proxy liveness. CPs could be implemented using any transport protocol (currently, several are available) with its own set of features such as compression, authentication, encryption, multicast, etc. Furthermore, each server or client can communicate with the DVS-SP with its own kind of CP and its own features.

The DVS-SP behavior is controlled by a startup configuration file. In this file, parameters such as the *low-water* and *high-water* load levels are specified. These values, together with the measurement period are sent to all server agents.

Two parameters control the scaling behavior, *min-servers* which set the minimum number of servers that must be active, and *max-servers* which indicate the maximum. If during a period with the minimum of active servers one or more server crashes, the DVS-SP starts new servers sequentially until reaching the *min-servers* value. It uses sequentially start-ups because the all servers could be restarted or it could be a temporary network failure.

The DVS-SP distinguishes between two types of server nodes. Those which are manually started and finished, and those which their start-up and shutdown is controlled by the DVS-SP, for the latter, the commands that should be used by the DVS-SP are also set in the configuration file.

The complete topology is described in the configuration file, Client nodes, Server nodes and services with their own parameter settings. Although and automatic configuration could be implemented which can discover about them, it is more secure that the DVS-SP knows the topology reading a local configuration file. Future works in which communication proxies and GCS provide strong authentication and encryption could offer these features.

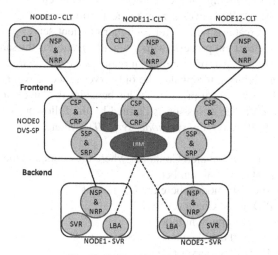

Fig. 3. DVS-SP architecture.

The reader should consider that this architecture was not designed to serve user applications such as web browsers as clients. It should be used among application services, i.e. web servers (Clients) which need to read/write files from/to network filesystems (Servers) or Database servers.

4.1 Main Service Proxy (MSP)

As was mentioned earlier, the MSP reads the configuration file which describes the cluster, initializes all data structures, and starts the other component threads.

In the configuration file four types of items are specified:

a. *MSP*: specifies the node name where the MSP runs, the node ID, the high-water and low-water load levels and the sampling period to calculate the servers' node load average.
b. *Server*: specifies the server node name and its node ID, the transport protocol its use and their features (such as compression, message batching, etc.). If the specified protocol is TCP or UDP, the sender and receiver port must be configured. If the server node is a VM, the commands and VM image could be specified to start the server on demand.

c. *Client*: specifies the client node name and its node ID, the transport protocol its use and their features (such as compression, message batching, etc.). If the specified protocol is TCP or UDP, the sender and receiver port must be configured.
d. *Services*: describes the name of the service (i.e. a fileserver), the external endpoint (*ext_ep*) in which the SP will receive requests from clients, the lowest (*low_ep*) and highest (*high_ep*) endpoint numbers which servers could use to serve the requests, and eventually the pathname of a server program to run on a server node on demand.

Services could be running on the server nodes (persistent service) or they could be started when the MSP receives a new request from a client (ephemeral service). Therefore, the MSP creates a *Session* for each pair of client-server processes. Once the MSP detects that a new client process (with a new PID) requests the same service on the same node using the same pair of endpoints, it removes the old *Session* from its database and, for ephemeral services, terminates the old server process. Afterward, it creates a new *Session* for the new pair of processes. This behavior is a piece of the auto-scaling mechanism of the DVS-SP. A session is defined by the following seven-tuple: {*dcid, clt_ep; clt_node, clt_PID, svr_ep, svr_node, svr_PID*}, where:

– *dcid*: is the ID of the Distributed Container where the client and the server run.
– *clt_ep*: is the Client endpoint number.
– *clt_node*: is the ID of the node where the Client runs.
– *clt_PID*: is the PID of the Client process.
– *svr_ep*: is the Server endpoint number.
– *svr_node*: is the ID of the node where the Server runs.
– *svr_PID*: is the PID of the Server process.

As an endpoint can be allocated by a process and then, after process termination, it can be reused by another process, the process' PID is the way in which the DVS-SP can distinguish among expired sessions. An expired session is that in which all values of the session matches but *clt_PID* or *svr_PID*.

As the LBM manages the load database of all Servers, the DVS-SP scheduling module can decide when it is time to allocate another Server node VM for new sessions (scale up) or when it is time to shut down one of them (scale down).

4.2 Client Receiver Proxy (CRP) and Server Sender Proxy (SSP)

When a new session starts, the client sends a message to the external endpoint (*ext_ep*) of the DVS-SP through its NSP to the CRP. The CRP compares several fields of the session database to find an active one that matches. If it does not exist, it searches the server's database for the first non-saturated server node (server load < *high-water*) and allocates it for that session. The reader might ask: why not choose the server with the least load? That policy would go against auto-scaling because an unloaded server quickly could acquire more workload and could not be removed from the cluster (scale down). If a program is specified for that service, the CRP sends a remote command to the server's node to execute the service server program. Then, the proxy message is forwarded to the server NRP.

When the CRP receives a new request from a client, and all active servers are saturated or they have no free endpoints to attend new requests, the CRP can start a VM of a new server node.

4.3 Server Receiver Proxy (SRP) and Client Sender Proxy (CSP)

When an SRP receives a message from its server's NSP it checks if a session exists. If all the session's parameters match, but the server's PID, it removes it as an expired session. If a program was specified for that service, it sends a remote command to the server's node to terminate the current server process. If the server's PID also matches, it gets the client endpoint field of the session and queues the proxy message into the CSP message queue. As the CSP is waiting for messages in its queue, it forwards the message to its Client NRP.

Several endpoint and node conversions are done in the header of proxy messages on CRP and SRP to hide the real architecture from clients and servers. Clients only request the DVS-SP as their single server, and servers only reply to the DVS-SP as their single client (service proxy behavior).

4.4 Load Balancer Monitor (LBM)

The LBM collects information about the load level of server nodes. The Load Balancer Agents (LBA), report their node load levels only when they change. The load levels are defined as *LVL_IDLE, LVL_LOADED,* and *LVL_SATURATED.* These values are calculated using configuration parameters specified in the configuration file.

The LBM manages and keeps updated the node status database used by CRPs to allocate servers for new sessions. When server nodes fail or a network partition occurs (reported by the GCS), the LBM deletes all the sessions with the faulty or unreachable nodes.

If the load level of all active servers is *LVL_SATURATED* during a specified *start_vm_period*, the LBM orders the hypervisor to start a new VM (scale up) as a server node. Similarly, if a server node has no active sessions during a specified *shutdown_vm_period*, the LBM orders the hypervisor to shut down its VM (scale down). Both, *start_vm_period* and *shutdown_vm_period* are parameters specified in the configuration file.

4.5 Load Balancer Agents (LBA)

Each LBA periodically evaluates the load of its node. Currently, the load of a node is defined as the mean of CPU usage (reported by Linux in the pseudo-file */proc/stat*) in a *lba_period*. This value is specified in the DVS-SP configuration file and sent by the LBM in a multicast message to all servers. Although there are additional metrics that could be considered to describe the load of a node [16], such as memory usage, network traffic, disk I/O, etc., only CPU usage was considered to simplicity the prototype implementation.

Each server node keeps a *load_lvl* variable that stores load level in the last period. At the start of each period, if the new load level value differs from *load_lvl*, the LBA reports this new level to the LBM using the GCS, and updates *load_lvl*. Therefore, the dissemination of load information is event-driven. This mechanism consumes lower network bandwidth than a periodic one [17]. In the case that the LBM is doesn't alive or is unreachable, the LBA doesn't report any message. Then, when LBM comes back, the LBA starts to report the load level again.

4.6 Membership and Fault Tolerance

As was mentioned earlier, the DVS-SP prototype uses Spread Toolkit as its GCS, but it also uses it as its failure detector. Spread detects the *join* and the *leave* of new nodes as the basic membership operations. But, it also detects host and network faults.

If the LBA fails, the GCS reports a disconnected member. If the server node fails, it reports a network partition, although the cause could have been the disconnection of the node because it hasn't got a way to distinguish a network failure from a node failure. On a server node failure, the DVS-SP closes all active sessions with the faulty server node and reports an EDVSDSTDIED (destination endpoint just died) error to the clients. On a client node failure, the DVS-SP closes all active sessions with the faulty client node and reports an EDVSSRCDIED (source endpoint just died) error to the servers.

On a DVS-SP failure, clients and servers don't have to do nothing, except waiting for that the DVS-SP will be restored from the fault, for this reason, a replicated DVS-SP is proposed as a future work.

5 Evaluation

This section describes the tests, benchmarks and micro-benchmarks used to verify the correct operation of the DVS-SP in a DVS virtual cluster. It should be considered that the tests should have been carried out in a home virtualized environment and not a physical environment as a consequence of the inability to access the laboratories during 2020 and 2021 due to the regulations established by the national government in relation to COVID-19. This fact does not imply important consequences to demonstrate the correct behavior of the DVS-SP, but for performance measurements. It's known that CPU, memory, disk, network virtualization could distort the results.

The hardware used to perform the tests was a PC with a 6-core/12-threads AMD Ryzen 5 5600X CPU, 16 GB of RAM, SSD and SATA disks. The virtualization was carried out using VMware Workstation version 15.5.0 running on Windows 10 and a cluster of 6 nodes was configured, each node in a VM: NODE{0–5}. Each VM was assigned a vCPU and 1 GB of RAM. The VMs were clones of each other running Linux kernel 4.9.88 modified with the DVK module. The DVS-SP runs on NODE0; servers run on NODE{1,2}, and clients run on NODE{3–5}. This virtual cluster was used to test the correct behavior of the DVS-SP on allocating new sessions to new servers when the other servers are saturated and, exchange load information among the LBAs and the LBM and to test fault-tolerance on server crashes, node failures, or network partitions.

The following are measurements obtained with the *lmbench* [18] tools that allow knowing the performance of the used infrastructure:

– *lat_tcp* (TCP client-server latency between 2 nodes): 557.88 [μs]
– *bw_tcp* (TCP client-server bandwidth between 2 nodes): 186.38 [MB/s]

To evaluate the DVS-SP performance (taking into account the previously mentioned test environment) a minimal cluster of 3 nodes was used: DVS-SP run in NODE0, NODE1 was the server node, and NODE2 was client node. Two main metrics were measured:

- *Single Session Latency*: Two programs were used, a latency client and a latency server. The server program waits for a request and, when it receives one, it replies to the client. The client sends a request and then it waits for the reply, measuring the elapsed time between these two events. The message transfer throughput of this pair of processes is also measured.
- *Single Session Data transfer throughput*: A file transfer pair of programs was used. The client can request a GET operation to transfer a file from server to client, or a PUT operation to transfer a file from client to server. The server measures the time between the first request received from the client and the last message sent to it, then, it calculates the throughput.

In Fig. 4(A), the results of latency tests are presented, where:

– *Local HTTPping*: The client sends an HTTPing to a web server on the same node.
– *Local M3-IPC Latency*: The client and server M3-IPC programs were executed on the same node.
– *Remote HTTPping*: The client sends an HTTPing to a web server on another node.
– *Remote M3-IPC Latency:* The client latency program was run on one node and the server latency program was executed on another node, both using M3-IPC with their nodes connected by communication proxies.
– *Proxy Through HTTPping Latency*: The communications between the client and the server traverses an HTTP proxy (*simpleproxy* [19]).
– *Proxy Through M3-IPC Latency*: The communications between the client and the server traverses the DVS-SP.

The penalty in latency is about 4% among remote and DVS-SP through communications.
In Fig. 4(B), the results of data transfer throughput tests are presented where, Remote means that the client and the server are communicated directly through the network and Proxy means that they communicate with an HTTP proxy or DVS-SP proxy between them. The notation on Fig. 4(B) is:

– *WGET*: The client uses *wget* to transfer a file from a web server.
– *M3-GET:* An M3-IPC client gets a file from an M3-IPC server.
– *M3-PUT:* An M3-IPC client sends a file to an M3-IPC server.

Several tests of file transfers were done using file sizes of 1 Mbyte, 10 Mbytes and 100 Mbytes with *simpleproxy* [19] and *nginx* [4] as proxies. The data transfer throughput is reduced between 52%–55% using the DVS-SP against a direct transfer between client and server. The delay of copying metadata between the client proxy and the server proxy introduces a significant increase in latency. Currently, the DVS-SP prototype uses a dedicated Linux thread for each proxy (senders and receivers as it is shown in Fig. 3). Therefore, a proxy receiver transfers metadata to a proxy sender through an XSI message queue, allowing the receiver to quickly be ready to receive new messages. This design decision also requires intensive use of mutexes and condition variables system calls. Ongoing work changes this behavior, so the receiver proxy also sends messages to the destination node avoiding the context switching between threads, the use of message queues and lower use of synchronization and semaphores system calls.

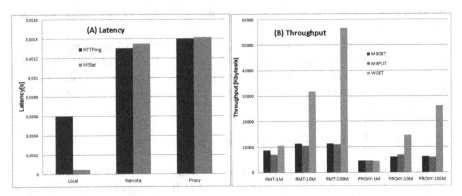

Fig. 4. (A) Latency and (B) Data transfer throughput.

As previously mentioned, the DVS-SP supports the starting up of a dedicated server program (on a running server node) once the first client request arrives to the SP. The mean time measured to startup the server process was 30 [ms]. Once this single session finished, the DVS-SP orders the node agent to terminate this server process.

Another feature of the DVS-SP is to start a new server node VM when all running nodes are saturated or there are no free endpoints or sessions that can used. The mean time measured to start a server node VM was 30 [s]. During this period, the client process receives an EDVSAGAIN (resource temporarily unavailable) reply message from the DVS-SP. On the other hand, if a server node is idle during a specified period, the DVS-SP orders the hypervisor to shutdown it.

To evaluate the DVS-SP performance under higher network loads, several tests of concurrent file transfers (GET) of files of 100 Mbytes each was done. The server side virtual Ethernet interface of the DVS-SP registered a maximum outbound bandwidth by 307,67 [Mbit/s] (measured by *vnstat* [20]). The average CPU utilization (measured by *vmstat* [21]) during the transfers, is shown in Fig. 5.

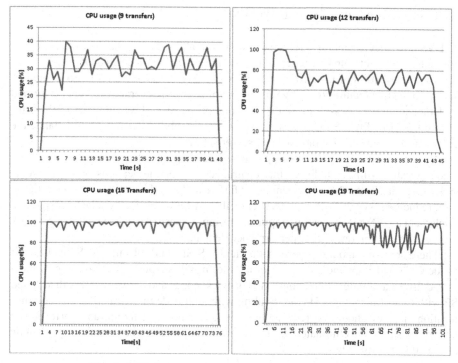

Fig. 5. DVS-SP CPU usage during concurrent file transfers.

Using these graphics, the average throughput can be estimated as in Table 1.

Table 1. DVS-SP average throughput.

# Concurrent transfers	Transfer time [s]	Throughput [Mbytes/s]
9	40	22,50
12	42	28,57
15	74	20,27
19	100	19,00

Currently, the DVS-SP proxies and the communication proxies of client and server nodes use the TCP protocol with compression enabled for data transfers. Previous comparative tests on the performance of M3-IPC communication proxies [22] showed that the TIPC protocol and a custom protocol based on raw Ethernet have better performance than TCP. The utilization of these protocols is left for future enhanced versions of the DVS-SP.

6   Conclusions and Future Works

A DVS provides scalability, reliability, and availability, it is simple to deploy and configure, and lightweight in terms of requirements which reduces the OPEX.

The contribution of this article is to present a Service Proxy with Load Balancing and Auto Scaling features for a DVS as a proof of concept. Several tests were performed to demonstrate the DVS-SP capabilities as one of the several features that DVS architecture has. This DVS-SP uses a centralized and probabilistic approach to balance the computational loads and avoid backend server saturation.

The use of an AP is a common practice deploying Cloud Applications. However, a configuration with a single AP that centralizes all communications between clients and servers has in the AP a single point of failure and a performance bottleneck; therefore, reducing service availability and scalability. A future project will be to design and implement a DVS-SP cluster with replication support that can handle nodes and network failures and can tolerate higher performance demands.

There are two ongoing projects about the DVS-SP: 1) Replacing the Spread toolkit as the GCS and implementing an internal one. It will support FIFO ordering in message delivery and failure detections for processes, nodes and networks. 2) A distributed SP with Load Balancing and Auto Scaling without a node in the middle. The communications latency should be reduced and the single point of failure and performance bottleneck should be eliminated because each client process will communicate with its allocated server directly.

References

1. Pessolani, P., Cortes, T., Tinetti, F., Gonnet, S.: An Architecture Model for a Distributed Virtualization System, Cloud Computing 2018, Barcelona, España (2018)
2. Martin Fowler: https://martinfowler.com/articles/microservices.html. Accessed Dec 2021
3. Pessolani, P., Taborda, M., Perino, F.: Service Proxy with Load Balancing and Auto Scaling for a Distributed Virtualization System, Memorias del Congreso Argentino en Ciencias de la Computación - CACIC 2021, pp. 480–489, December 2021. ISBN: 978-987-633-574-4
4. Nginx. https://www.nginx.com/. Accessed Dec 2021
5. HAproxy. http://www.haproxy.org/. Accessed Dec 2021
6. Varnish. http://varnish-cache.org/. Accessed Dec 2021
7. Traefik. https://traefik.io/traefik/. Accessed Dec 2021
8. Apache Traffic Server. https://trafficserver.apache.org/. Accessed Dec 2021
9. Hedstrom, L.: Apache Traffic Server: More Than Just a Proxy, talk at USENIX (2011)
10. ATS vulnerabilities. https://vuldb.com/?product.apache:traffic_server. Accessed Dec 2021
11. Pessolani, P., Cortes, T., Tinetti, F.G., Gonnet, S.: An IPC software layer for building a distributed virtualization system. In: CACIC 2017, La Plata, Argentina (2017)
12. Zookeeper. https://zookeeper.apache.org/. Accessed Dec 2021
13. Ongaro, D., Ousterhout, J.: In search of an understandable consensus algorithm. In: 2014 USENIX Annual Technical Conference (2014). ISBN 978-1-931971-10-2
14. The Spread Toolkit. http://www.spread.org. Accessed Dec 2021
15. Moser, L.E., et al.: Extended virtual synchrony. In: Proceedings of the IEEE 14th International Conference on Distributed Computing Systems, Poznan, Poland, June 1994

16. Ghiasvand, S., et al.: An analysis of MOSIX load balancing capabilities. In: International Conference on Advanced Engineering Computing and Applications in Sciences, November 20–25, Lisbon, Portugal (2011)
17. Beltrán, M., Guzmán, A.: How to balance the load on heterogeneous clusters. Int. J. High Perform. Comput. Appl. **23**(1), 99–118 (2009)
18. Lmbench. http://lmbench.sourceforge.net/. Accessed Dec 2021
19. Simpleproxy. https://manpages.debian.org/testing/simpleproxy/simpleproxy.1.en.html. Accessed Dec 2021
20. Vnstat. https://humdi.net/vnstat/. Accessed Dec 2021
21. Vmstat. https://linux.die.net/man/8/vmstat. Accessed Dec 2021
22. Gonzalez, R.: A Raw Ethernet Proxy for a Distributed Virtualization System (in Spanish), 5to Congreso Nacional de Ingeniería Informática/Sistemas de Información – CONAIISI 2017, Santa Fe, Argentina (2017)

Innovation in Software Systems

Ontology Metrics and Evolution in the GF Framework for Ontology-Based Data Access

Sergio Alejandro Gómez[1,2]([⊠]) and Pablo Rubén Fillottrani[1,2]

[1] Laboratorio de I+D en Ingeniería de Software y Sistemas de Información (LISSI),
Departamento de Ciencias e Ingeniería de la Computación,
Universidad Nacional del Sur, San Andrés 800, Bahía Blanca, Argentina
{sag,prf}@cs.uns.edu.ar
[2] Comisión de Investigaciones Científicas de la Provincia de Buenos Aires
(CIC-PBA), Calle 526 entre 10 y 11, La Plata, Argentina
https://lissi.cs.uns.edu.ar/

Abstract. There exists a need of producing high-quality ontologies for Semantic Web applications that access legacy, relational and non-relational data sources. We present GF, a tool for materialization of ontologies from relational and non-relational data sources including H2 databases, CSV files and Excel spreadsheets. We evaluate a sample case generated by GF with a third-party ontology evaluation tool called OntoMetrics. We also introduce an scripting language for performing Ontology-Based Data Access allowing a semi-naive user to automate ontology generation and document ontologies by adding annotations. The results obtained show that the ontologies generated with GF are reasonably good for being used in Semantic Web applications because they are validated correctly and pass all of the filters for the OWL2 main profiles, thus making them suitable for processing with lightweight reasoners. The metrics originally indicated that our application was lacking quality in the annotation area regarding the documentation of the classes and properties generated by the application and that the functionality introduced by the scripting language allows to generate correctly annotated ontologies. An executable standalone application along with the data used in this paper are uploaded to a GitHub repository for reproducibility of the results presented here.

Keywords: Ontologies · Ontology-based data access · Metrics · Knowledge representation

1 Introduction

In the Semantic Web (SW), data resources have precise meaning given in terms of ontologies making them apt to be machine processed. An ontology is a logical theory described in the OWL language whose factual information is described in

P. Pesado and G. Gil (Eds.): CACIC 2021, CCIS 1584, pp. 237–253, 2022.
https://doi.org/10.1007/978-3-031-05903-2_16

the RDF language. OWL/RDF ontologies are identified by an IRI and are composed of concepts/classes and relations/properties among classes and/or classes and values. This seemingly simplification allows improved efficiency in reasoning for which ontologies comprise an amenable tool for managing legacy data.

Ontology-based data access (OBDA) is concerned with enriching both relational and non-relational, usually legacy, data sources with an ontology describing the model of an application domain. One common approach to OBDA is materialization where data sources are translated into ontologies (both at schema and instance information levels) and then enriched with application data. We have developed a framework, called GF, for performing OBDA with legacy data sources based on materialization, that is currently in a prototypical state but to which we are incrementally adding functionality following a problem-solving driven approach to information interoperability. The current implementation of GF allows to deal with H2 databases, CSV files and Excel spreadsheets.

Evaluation of the ontologies produced with our framework is important. The process of ontology evaluation is concerned with ascertain quality and correctness of ontologies. This ultimately allows quantifying the suitability of a given ontology to the purposes for which it has been built. Several metrics have been proposed for ontologies over the last years (see Sect. 3 and related work literature [1–7] for details) showing the importance of the field. A notable third-party project that accrues the most part of the ontology metrics developed is Ontometrics that introduces itself to the final user as a web application allowing to obtain meaningful metrics from an ontology loaded as a text file.

In this paper, we revisit a case study on solving the construction of a university library ontology from a set of legacy data sources using GF [8]. We then evaluate the obtained ontology with a third-party ontology evaluation tool called OntoMetrics. We thus obtained several descriptors that we have used to debug our implementation and that indicated that the prototype was working correctly but needed some improvements. The results obtained show that the ontologies generated with GF are reasonably good for being used in SW applications as they are validated correctly and pass all of the filters for the OWL2 DL, OWL2 EL, OWL2 QL and OWL RL profiles, thus making them suitable for processing with lightweight reasoners. The metrics indicated that our application was lacking quality in the annotation area regarding the documentation of the classes and properties generated by the application. This pitfall motivated further improvements that are discussed in this extended version.

This paper extends and consolidates results presented in [9]. In particular, the present version of the article improves those results by introducing a scripting language that allows to automate the OBDA workflow that it is usually done via the GUI in the GF framework. Besides a new functionality implemented via the GUI and the scripting language was added in order to introduce annotations in the ontology objects to solve the pitfalls described above in regards to the lack of impact in the annotation metrics. We therefore re-ran the annotation metrics to test the viability of our approach with positive results. An executable standalone application along with the data used in this paper are uploaded to a GitHub repository[1] for reproducibility of the results presented here.

The rest of the article is structured as follows. In Sect. 2, we review the basic functionality provided by the GF framework application. In Sect. 3, we review some of the most important approaches to metrics used for ontologies. In Sect. 4, we present the experiments that we performed to measure ontology metrics on ontologies produced by our application and the properties emerging from the cases that we observed. In Sect. 5, we present the scripting language for OBDA and ontology evolution. Finally, in Sect. 6, we present our conclusions and foresee future work.

2 Ontology-Based Data Access in GF

The GF framework allows to materialize OWL/RDF ontologies from heterogeneous, legacy, both relational and non-relational data sources. We explained its functionality in previous work (see [8] and references therein). Assume a relational database (RDB) as in Fig. 1, a typical workflow with GF to produce a working ontology as in Fig. 2 from such database along with Excel and CSV files would typically be: (1) Create a new ontology with IRI http://foo.org/ and file name OntoLibrary.owl. (2) Establish a connection to the RDB Users-Theses-Loans.db. (3) Materialize an initial ontology with the DB contents. (4) Build intermediate classes Material, Printed, with their respective attributes. (5) Establish that Thesis and Printed are subclasses of Material. (6) Modify schema expressing that Loan is now related to Material instead of Thesis by changing the range of property id and http://foo.org/Loan/ref-id from, select ref-id as object property name and *edit object property*), selecting http://foo.org/Material as new class range. (7) Create class PostgradThesis as a subclass of Thesis. (8) Create classes StudentUser, TeacherUser, GraduateThesis, MScThesis, PhDThesis with the SQL filters by introducing them directly or visually building them:

- StudentUser: SELECT "User"."userID" FROM "User" WHERE "User"."type"='S'
- TeacherUser: SELECT "User"."userID" FROM "User" WHERE "User"."type"='T'
- GraduateThesis: SELECT "Thesis"."id" FROM "Thesis" WHERE "Thesis"."type"='S'
- MScThesis: SELECT "Thesis"."id" FROM "Thesis" WHERE "Thesis"."type"='M'
- PhDThesis: SELECT "Thesis"."id" FROM "Thesis" WHERE "Thesis"."type"='D'

This also establishes that StudentUser and TeacherUser are subclasses of User, GraduateThesis is a subclass of Thesis, and both MScThesis and PhDThesis are subclasses of Thesis (which is made automatically by the system and we will later see that this option has to actually be optional). (9) Establish that PhDThesis and MScThesis are subclasses of PostgradThesis. (10) Establish disjoint classes MScThesis and PhDThesis. (11) Create class Magazine using Excel schema file magazines.xsc and Excel file Magazines.xlsx. (12) Establish Magazine as subclass of Printed. (13) Create class Book using CSV schema file books.sch and CSV file books.sch. (14) Establish Book as subclass of Printed. (15) Finally, save the ontology.

---

[1] See https://github.com/sergio-alejandro-gomez/gfobda.

240   S. A. Gómez and P. R. Fillottrani

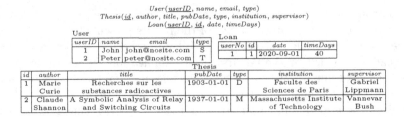

$User(\underline{userID},\ name,\ email,\ type)$
$Thesis(\underline{id},\ author,\ title,\ pubDate,\ type,\ institution,\ supervisor)$
$Loan(\underline{userID},\ \underline{id},\ date,\ timeDays)$

User

userID	name	email	type
1	John	john@nosite.com	S
2	Peter	peter@nosite.com	T

Loan

userNo	id	date	timeDays
1	1	2020-09-01	40

Thesis

id	author	title	pubDate	type	institution	supervisor
1	Marie Curie	Recherches sur les substances radioactives	1903-01-01	D	Faculte des Sciences de Paris	Gabriel Lippmann
2	Claude Shannon	A Symbolic Analysis of Relay and Switching Circuits	1937-01-01	M	Massachusetts Institute of Technology	Vannevar Bush

Fig. 1. Relational instance of the library's database concerning Users, Theses and Loans

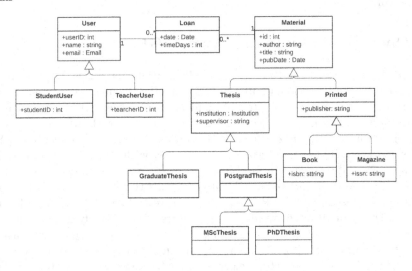

Fig. 2. Ontology for the university library

The validation process of the obtained ontology[2] determined that it complies with OWL 2 DL, OWL 2 EL, OWL 2 QL, and OWL 2 RL profiles. But a visualization of the ontology[3] shows that there are redundant *is-a* relations (e.g. PhDThesis is a subclass of both PostgradThesis and Thesis, where the latter is redundant as it does not need to be explicitly expressed and could be determined by an OWL reasoner). See Fig. 3.

---

[2] An online validator can be found at http://visualdataweb.de/validator/.

[3] OWL/RDF ontologies can be visualized with http://www.visualdataweb.de/webvowl/#.

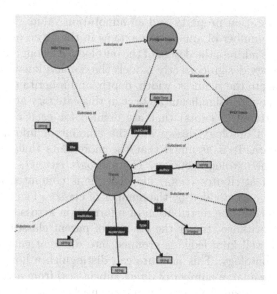

Fig. 3. Part of the ontology generated with GF

3   Measuring Ontologies with *OntoMetrics*

Accountability is an integral part to the success of any endeavor. Accountability requires metrics that truly reflect the desired outcomes of a program. Metrics are measures of quantitative assessment commonly used for assessing, comparing, and tracking performance or production. *OntoMetrics* is a third-party web-based tool that validates and displays statistics about a given ontology. Ontometrics implements several measurement frameworks such as [7,10]. We now give an overview of the metrics computed by Ontometrics that we used on the experiments reported in Sect. 4. For further details, we refer the reader to the Ontometrics web page[4].

Base metrics include the counting of classes, axioms, objects, etc. that show the quantity of ontology elements. *Class axioms metrics* count the number of subclass, equivalent classes and disjoint classes relations. *GCI* counts the number of the General Concept Inclusion (GCI). *HiddenGCI* counts the number of hidden GCIs in an ontology imports closure. A GCI is regarded to be a hidden GCI if it is essentially introduce via an equivalent class axiom and a subclass axioms where the LHS of the subclass axiom is named. For example, *A equivalentTo p some C, A subClassOf B* results in a hidden GCI. *Object property axiom metrics* and *Data property axioms metrics* quantify the presence of object properties and data properties, resp. *Individual axioms metrics* measure are axioms concerning individuals in the extension of classes and properties.

Annotations can be used to associate information to ontologies, this information could be the version of the ontology or the creator. The annotation itself

---

[4] See https://ontometrics.informatik.uni-rostock.de/ontologymetrics/.

consists of an annotation property and an annotation value. *Annotation axiom metrics* count the number of annotation axioms in the given ontology.

Schema metrics address the design of the ontology. Although it is not possible to tell if the ontology design correctly models the domain knowledge, metrics in this category indicate the richness, width, depth, and inheritance of an ontology schema design. The most significant metrics in this category are described next. The number of attributes (slots) that are defined for each class can indicate both the quality of ontology design and the amount of information pertaining to instance data. It is assumed that the more slots that are defined, the more knowledge the ontology conveys. The *attribute richness* is defined as the average number of attributes (slots) per class, it is computed as the number attributes for all classes divided by the number of classes. The *inheritance richness* measure describes the distribution of information across different levels of the ontology's inheritance tree or the fan-out of parent classes. This is a good indication of how well knowledge is grouped into different categories and subcategories in the ontology. This measure can distinguish a horizontal ontology (where classes have a large number of direct subclasses) from a vertical ontology (where classes have a small number of direct subclasses). An ontology with a low inheritance richness would be of a deep (or vertical) ontology, which indicates that the ontology covers a specific domain in a detailed manner, while an ontology with a high IR would be a shallow (or horizontal) ontology, which indicates that the ontology represents a wide range of general knowledge with a low level of detail. The *relationship richness* metric reflects the diversity of the types of relations in the ontology. An ontology that contains only inheritance relationships usually conveys less information than an ontology that contains a diverse set of relationships. The relationship richness is represented as the percentage of the (non-inheritance) relationships between classes compared to all of the possible connections that can include inheritance and non-inheritance relationships. The relationship richness of a schema is defined as the ratio of the number of (non-inheritance) relationships (P), divided by the total number of relationships defined in the schema (the sum of the number of inheritance relationships (H) and non-inheritance relationships (P)). The following relationships are being counted as non-inherited relationships: object properties, equivalent classes, and disjoint classes. The subclasses are being handled as inheritance relationships [7,10]. The *attribute-class ratio* metric represents the relation between the classes containing attributes and all classes. The *equivalence ratio* calculates the ratio between similar classes and all classes in the ontology. The *axiom class ratio* metric describes the ratio between axioms and classes. It is calculated as the average amount of axioms per class. The *inverse relations ratio* metric describes the ratio between the inverse relations and all relations. The *class relation ratio* describes the ratio between the classes and the relations in the ontology [7,10].

The manner in which data is located inside an ontology is an important measure of the quality of the ontology as it indicates the effectiveness of the ontology design as well as the amount of real-world knowledge represented by the ontology[5]. *Instance metrics* include metrics that describe the knowledge-base as a whole, and metrics that describe the way each schema class is being utilized in the knowledgebase. These are collectively known as *knowledgebase metrics*. The *average population* (i.e. the average distribution of instances across all classes) measure is an indication of the number of instances compared to the number of classes. It can be useful if the ontology developer is not sure if enough instances were extracted compared to the number of classes. Formally, the average population (AP) of classes in a knowledgebase is defined as the number of instances of the knowledgebase (I) divided by the number of classes defined in the ontology schema (C). The result will be a real number that shows how well is the data extraction process that was performed to populate the knowledge-base. For example, if the average number of instances per class is low, when read in conjunction with the previous metric, this number would indicate that the instances extracted into the knowledgebase might be insufficient to represent all of the knowledge in the schema. Notice that some of the schema classes might have a very low number or a very high number by the nature of what it is representing. The *class richness* metric is related to how instances are distributed across classes. The number of classes that have instances in the knowledgebase is compared with the total number of classes, giving a general idea of how well the knowledgebase utilizes the knowledge modeled by the schema classes. Thus, if the knowledgebase has a very low class richness, then the knowledgebase does not have data that exemplifies all the class knowledge that exists in the schema. On the other hand, a knowledgebase that has a very high class richness indicates that the data in the knowledgebase represents most of the knowledge in the schema. The class richness (CR) of a knowledgebase is defined as the percentage of the number of non-empty classes (classes with instances) (C') divided by the total number of classes (C) defined in the ontology schema.

Class metrics examine the classes and relationships of ontologies. The *class connectivity* metric is intended to provide an indication of what classes are central in the ontology based on the instance relationship graph (where vertices represent instances and edges represent the relationships between them). This measure works in tandem with the importance metric to create a better understanding of how focal some classes function. This measure can be used to understand the nature of the ontology by indicating which classes play a central role compared to other classes. The *connectivity of a class* is defined as the total number of relationships that instances of the class have with instances of other classes. The *class importance* metric calculates the percentage of instances that belong to classes at the inheritance subtree rooted at the current class with respect to the total number of instances. This metric is important because it helps in individualizing which areas of the schema are in focus when the instances are added to the knowledgebase. Although this measure does not consider the domain characteristics, it can still be used to give an idea on what parts of the ontology are considered focal and what parts are on the edges. The *importance*

---

[5] See https://ontometrics.informatik.uni-rostock.de/wiki/index.php/Knowledgebase_Metrics.

of a class is defined as the percentage of the number of instances that belong to the inheritance subtree rooted at in the knowledgebase compared to the total number of class instances in the knowledgebase.

The *class inheritance richness* measures details the schema IR metric mentioned in schema metrics and describes the distribution of information in the current class subtree per class. This measure is a good indication of how well knowledge is grouped into different categories and subcategories under this class. Formally, the inheritance richness (IRc) of class C_i is defined as the average number of subclasses per class in the subtree. The number of subclasses for a class C_i is defined as $|H^C(C_1, C_i)|$ and the number of nodes in the subtree is $|C'|$. The result of the formula is a real number representing the average number of classes per schema level. The interpretation of the results of this metric depends highly on the nature of the ontology. Classes in an ontology that represents a very specific domain will have low IRc values, while classes in an ontology that represents a wide domain will usually have higher IRc values. The *class readability* metric indicates the existence of human readable descriptions in the ontology, such as comments, labels, or captions. This metric can be a good indication if the ontology is going to be queried and the results listed to users. Formally, the readability of a class is defined as the sum of the number of attributes that have comments and the number of attributes that are labels the class has. The result of the formula is an integer representing the availability of human-readable information for the instances of the current class. The *class relationship richness* is a metric reflecting how much of the relationships defined for the class in the schema are actually being used at the instance level. This is another good indication of the utilization of the knowledge modeled in the schema. The *class children* is a count-metric that measures the number of immediate descendants of a given class, also known as a number of children [10]. The *class instances* metric displays the number of instances of a given class. *Class properties* metrics summarize the properties of a given class [7].

Graph or structural metrics calculate the structure of ontologies. *Cardinality* is a property of graphs which expresses a graph related number of specific elements. *Absolute root cardinality* is a property of a directed graph which represents the number of root nodes of the graph. *Absolute root cardinality* is a property of a directed graph which is related to leaf node sets and represents the number of leaf nodes of the graph. *Absolute sibling cardinality* is a property of a directed graph which is related to sibling node sets and represents the number of sibling nodes of the graph. *Depth* is a property of graphs which is related to cardinality of paths existing in the graph. The arcs which are considered are only *is-a* arcs but this only applies to directed graphs.

4   Experimental Results

As explained above, *OntoMetrics* is a web-based tool that validates and displays statistics about a given ontology, where a user can upload an OWL/RDF ontology source. Using this third-party application, we conducted several experiments for applying the metrics described in Sect. 3 to ontologies produced with the GF application. Here, we, in particular, present the results obtained in relation to the library example ontology described in Sect. 2.

The results of computing base and class-axioms metrics on the library ontology from Fig. 2 are shown in Table 1. The results of computing Object and data property axioms metrics on the library ontology from Fig. 2 are shown in Table 2. The individual and annotation axioms metrics on the library ontology are presented in Table 3. The schema, knowledgebase and graph metrics are presented in Table 4. Finally, the class metrics computed for each of the classes of the library case study are shown in Table 5.

Table 1. Base and class axioms metrics of the library ontology in Fig. 2

Base metrics	
Axioms:	208
Logical axioms count:	150
Class count:	13
Total classes count:	13
Object property count:	2
Total object properties count:	2
Data property count:	35
Total data properties count:	35
Properties count:	37
Individual count:	8
Total individuals count:	8
DL expressivity:	$\mathcal{ALC}(D)$

Class axioms:	
SubClassOf axioms count:	12
Equivalent classes axioms count:	0
Disjoint classes axioms count:	1
GCICount:	0
HiddenGCICount:	0

Table 2. Object and data property axioms metrics of the library ontology in Fig. 2

Object property axioms	
SubObjectPropertyOf axioms count:	0
Equivalent object properties axioms count:	0
Inverse object properties axioms count:	0
Disjoint object properties axioms count:	0
Functional object properties axioms count:	0
Inverse functional object properties axioms count:	0
Transitive object property axioms count:	0
Symmetric object property axioms count:	0
Asymmetric object property axioms count:	0
Reflexive object property axioms count:	0
Irreflexive object property axioms count:	0
Object property domain axioms count:	2
Object property range axioms count:	2
SubPropertyChainOf axioms count:	0

Data property axioms	
SubDataPropertyOf axioms count:	0
Equivalent data properties axioms count:	0
Disjoint data properties axioms count:	0
Functional data property axioms count:	0
Data property domain axioms count:	35
Data Property range axioms count:	35

Table 3. Individual and annotation axioms metrics of the library ontology in Fig. 2

Individual axioms	
Class assertion axioms count:	12
Object property assertion axioms count:	2
Data property assertion axioms count:	49
Negative object property assertion axioms count:	0
Negative data property assertion axioms count:	0
Same individuals axioms count:	0
Different individuals axioms count:	0

Annotation axioms	
Annotation axioms count:	0
Annotation assertion axioms count:	0
Annotation property domain axioms count:	0
Annotation property range axioms count:	0

Table 4. Schema, knowledgebase and graph metrics of the library ontology in Fig. 2

Schema metrics	
Attribute richness:	2.692308
Inheritance richness:	0.923077
Relationship richness:	0.2
Attribute class ratio:	0.0
Equivalence ratio:	0.0
Axiom/class ratio:	16.0
Inverse relations ratio:	0.0
Class/relation ratio:	0.866667

Knowledgebase metrics	
Average population:	0.615385
Class richness:	0.692308

Graph metrics	
Absolute root cardinality:	3
Absolute leaf cardinality:	8
Absolute sibling cardinality:	13
Absolute depth:	37
Average depth:	2.466667
Maximal depth:	4
Absolute breadth:	15
Average breadth:	2.5
Maximal breadth:	4
Ratio of leaf fan-outness:	0.615385
Ratio of sibling fan-outness:	1.0
Tangledness:	0.153846
Total number of paths:	15
Average number of paths:	3.75

Table 5. Class metrics of the library ontology in Fig. 2

Class metrics	Book	GraduateThesis	Loan	MScThesis	Magazine	Material	PhDThesis	PostgradThesis	Printed	StudentUser	TeacherUser	Thesis	User
Class connectivity:	0	0	2	0	0	0	0	0	0	0	0	0	0
Class fulness:	0.0	0.0	0.0	0.0	0.0	0.0	0.0	0.0	0.0	0.0	0.0	0.0	0.0
Class importance:	0.125	0.0	0.125	0.125	0.25	0.125	0.125	0.25	0.375	0.125	0.125	0.75	0.5
Class inheritance richness:	0.0	0.0	0.0	0.0	0.0	1.3	0.0	6.5	6.5	0.0	0.0	2.166	6.5
Class readability:	0	0	0	0	0	0	0	0	0	0	0	0	0
Class relationship richness:	0	0	0	0	0	0	0	0	0	0	0	0	0
Class children count:	0	0	0	0	0	10	0	2	2	0	0	6	2
Class instances count:	1	0	1	1	2	9	1	2	3	1	1	6	4
Class properties count:	0	0	0	0	0	0	0	0	0	0	0	0	0

Now we provide an interpretation of the results that we obtained. The class count coincides with the number of classes that we can manually count in the UML diagram of Fig. 2. The data property count exceeds the number of attributes seen in the UML diagram because every attribute of the database triggers the creation of one data property but in order to model the class hierarchy we have to manually create the attributes of each of the corresponding superclasses thus producing some overlap and consequently some redundancy too. The new functionality of modifying relations (see step (6) in the workflow described in Sect. 2) allows to continue having only 2 object properties (where one remained from Loan to User but another one that moved from Loan to Thesis towards Loan to Material, making a loan more generic than before). The number of subclass of axioms count (12) exceeds the expected number (10) from the UML diagram in Fig. (2.a) because GF produces redundant is-a relationships as it can be seen in Fig. (2.b)—there are two paths from MScThesis to Thesis and another two from PhDThesis to Thesis instead of one in each case. The results of Table 2 show that indeed the ontologies produced by GF have only property axioms defining attributes and dispense with other type of richer semantic relations that would impose an excessive time complexity worst-case on the reasoner, thus adhering to the tenets of the OBDA paradigm that propose only using light-weight ontologies to guarantee polynomial time complexity reasoning. Table 3 shows that only 12 of 13 classes have individuals, that is consistent with our case study where there are no StudentUser of the library. One important aspect revealed by the annotation axioms metrics is that GF is not impacting them at all, meaning that the usability of the ontologies produced would be at stake and pointing out a place of the framework where to introduce further improvement. In Table 4, we see that 3 roots are detected (User, Loan and Material) and 8 leaves, that is all consistent with the UML diagram in Fig. (2.b). According to Table 5, Thesis appears to be the most important class followed by User, results that coincide with the richness of the relational database diagram.

5   A Scripting Language for OBDA and Ontology Evolution

Our hypothesis is that a scripting language can be used to expedite the process of bootstrapping ontologies from several diverse data sources. Our research method consists of building an executable prototype to test our hypothesis. This approach improves our current work in that until now the user could perform bootstrapping of ontologies by materializing the ontology from the data sources by using GUI controls only, a process that despite its simplicity has to be redone every time the bootstrapping needs to be done what can be cumbersome and error prone. It addresses the problem of lacking annotations in ontologies. The limitation of the approach described here is that (i) every command is executed in isolation although many might be dependent on previous commands; (ii) the language relies on an English speaking user; (iii) despite the scripting language simplicity, some basic experience in programming is needed to use it effectively; (iv) there is no part of the scripting language for defining SQL mappings.

Here, we will explain the syntax and semantics of the scripting language for OBDA and ontology evolution.

Ontology Creation. The command create-ontology iri:*i* name:*n* filename:*f* creates a new (empty) ontology with IRI *i* and name *n*, that will be saved on file named *f*, that has to be a full path.

DB Connection. The command connect-to-db dbPath:*d* user:*u* pass:*p* connects to an H2 database stored in the file *d* using user *u* and password *p*.

Ontology Materialization. The command materialize-ontology materializes all the contents of the current database to the ontology using the direct mapping specification and adding both Tbox and Abox axioms to the current ontology.

Class Creation. The command create-class class:*c* [iri:*i*] creates a class named *c* with IRI *i*. The IRI is optional and if it is not specified, the IRI of the ontology will be used. In the current implementation, this option is limited to specify the name of the class and the IRI assigned to the class is the concatenation of the ontology IRI and the name of the class *c*.

Attribute Addition and Edition. The command add-attribute class:*c* name:*n* type:*t* Adds a property (attribute of base type) to class *c* with name *n* and type *t*. The type has to be one of string, integer, float, double, boolean, date, time, timestamp. The command add-ref-attribute class:*c* name:*n* type:*c'* Adds a relation (attribute of class type) to class *c* with name *n* and type *c'*. The type has to be one of the existing classes in the ontology. The command modify-type-of-attribute class:*c* attribute:*a* former-type:*t* desired-type:*t'* changes the type of the attribute *a* in class *c* to from *t* to *t'*. For example, for changing the type of attribute ref-id in class Loan to class Material several computations have to be made by the system: first the absolute IRI http://foo.org/Loan/ref-id as the name of the attribute has to be computed; second, the replacement IRI has to be also computed, viz. http://foo.org/Material. This can come up as a limitation (in the sense that full IRIs cannot be specified as it could be done in e.g. SPARQL) however we think that this maintains the usability of the approach simpler from the perspective of the final user. The command modify-class-of-attribute former-class:*c* attribute:*a* desired-class:*c'* modifies the domain of the attribute *a* from class *c* to *c'*.

Subclassing and Class Relations. The command add-subclass-relation super-class:c sub-class:c' adds a subclass axiom to the ontology establishing that class c' is a subclass of class c. The classes c and c' have to be valid classes of the ontology. The command establish-disjoint-classes class-one:c class-two:c' establishes that classes c and c' are disjoint with respect to each another. For example, establish-disjoint-classes class-one:MScThesis class-two:PhDThesis establishes that MScThesis and PhDThesis are disjoint.

Ontology Mapping. The command define-mapping-class class:c map:e defines the assertional box contents of class c in terms of an SQL mapping expressed as the SQL expression e. The command define-mapping-property property:p map:e defines the assertional box contents of property p in terms of an SQL mapping expressed as the SQL expression e.

Loading and Saving Ontologies. The command save-ontology saves the current ontology while save-ontology-as filename:f saves the current ontology to a new file named f. This file name becomes the current file name. If f does not exist, it is created in the process. If f already exists, its contents are overwritten. The command load-ontology filename:f iri:i loads the ontology stored in the file f and assigns the IRI i to the current ontology. The loaded ontology becomes current ontology. If the file does not exist or is an invalid ontology, an error message is produced and the process is aborted.

Adding Annotations. Annotations can be used to associate information to ontologies, this information could be the version of the ontology or the creator. The annotation itself consists of an annotation property and an annotation value. The evaluation of annotation metrics of our previous approach to OBDA showed that the prototype was not impacting these metrics. Therefore, to overcome this deficiency, we added some functionality to the user interface for adding annotations. Likewise, we introduce three additional commands for adding annotations in scripting mode as shown next. The command add-annotation-to-class class:c annotation:[a] language:l adds an annotation a to class c indicating that the language of a is l. The command add-annotation-to-attribute class:c attribute:n annotation:[a] language:l adds an annotation a to attribute n of class c in language l. The command add-annotation-to-ontology annotation:[a] language:l adds an annotation a to the ontology in language l.

Example 1. The annotations commands in Fig. 4 show how to add English annotations to the User class, the attribute name of the same class and to the ontology. This has the effect of producing the OWL code shown in Fig. 5. By adding several annotations, that can be consulted in the GitHub repository of the application, we got an improvement in this regard by obtaining an annotation axioms count equal to 1, and annotation assertion axioms count equal to 29.

```
add-annotation-to-class class:User annotation:[It is a class for representing users of the library.] language:en
add-annotation-to-attribute class:User attribute:name annotation:[Represents the name of the user in the library.] language:en
add-annotation-to-ontology annotation:[Represents the library data.] language:en
```

Fig. 4. Commands for adding annotations to classes, attributes and ontologies

```
<owl:DatatypeProperty rdf:about="http://foo.org/User/name">
    <rdfs:domain rdf:resource="http://foo.org/User"/>
    <rdfs:range rdf:resource="http://www.w3.org/2001/XMLSchema#string"/>
    <rdfs:comment xml:lang="en">Represents the name of the user in the library.</rdfs:comment>
</owl:DatatypeProperty>
<owl:Class rdf:about="http://foo.org/Book">
    <rdfs:subClassOf rdf:resource="http://foo.org/Printed"/>
    <rdfs:comment xml:lang="en">It is a class for representing the books of the library.</rdfs:comment>
</owl:Class>
<rdf:Description rdf:about="http://foo.org/">
    <rdfs:comment xml:lang="en">Represents the library data.</rdfs:comment>
</rdf:Description>
```

Fig. 5. Portion of OWL code for annotations for attribute name of class User, class Book, and the ontology

Next, we present a worked example that shows how to the ontology scripting language introduced above allows a user to implement the workflow for OBDA of heterogeneous data sources presented in Sect. 2.

Example 2. In Fig. 6, we can see how the workflow for ontology integration introduced in Sect. 2 can be codified using the scripting language presented above. Line (1) creates the ontology with IRI http://foo.org/ and in file c:/OntologyFile.owl. Line (2) connects to the H2 database Base-Library-01 with user sa and empty password. Line (3) materializes the ontology of (2). Lines (5)–(10) create classes Material and Printed with their attributes. Lines (11)–(12) establish that Thesis and Printed are subclasses of Material. Line (13) modifies the type of the attribute ref-id of class Loan. Lines (14)–(20) create classes Post-gradThesis, StudentUser, TeacherUser, GraduateThesis, MScThesis and PhdThesis. Lines (21)–(25) establish the OBDA mappings for several classes. Lines (26)–(30) establish subclass-superclass and disjointness relationships between classes. Lines (31)–(33) performs OBDA of an Excel file for magazines. Lines (34)–(36) performs OBDA of a CSV file for books. Finally, line (37) saves the ontology consolidating it to disk.

1. create-ontology iri:http://foo.org/ name:onto1 filename:c:/OntoLibrary.owl

2. connect-to-db dbPath:c:/Users-Theses-Loans user:sa pass:

3. materialize-ontology

4. create-class class:Material

5. add-attribute class:Material name:id type:integer

6. add-attribute class:Material name:author type:string

7. add-attribute class:Material name:title type:string

8. add-attribute class:Material name:pubDate type:date

9. create-class class:Printed

10. add-attribute class:Printed name:publisher type:string

11. add-subclass-relation super-class:Material sub-class:Thesis

12. add-subclass-relation super-class:Material sub-class:Printed

13. modify-type-of-attribute class:Loan attribute:ref-id former-type:Thesis desired-type:Material

14. create-class class:PostgradThesis

15. add-subclass-relation super-class:Thesis sub-class:PostgradThesis

16. create-class class:StudentUser

17. create-class class:TeacherUser

18. create-class class:GraduateThesis

19. create-class class:MScThesis

20. create-class class:PhDThesis

21. define-mapping-class super-class:User class:StudentUser map:[SELECT "User"."userID" FROM "User" WHERE "User"."type"='S']

22. define-mapping-class super-class:User class:TeacherUser map:[SELECT "User"."userID" FROM "User" WHERE "User"."type"='T']

23. define-mapping-class super-class:Thesis class:GraduateThesis map:[SELECT "Thesis"."id" FROM "Thesis" WHERE "Thesis"."type"='S']

24. define-mapping-class super-class:PostgradThesis class:MScThesis map:[SELECT "Thesis"."id" FROM "Thesis" WHERE "Thesis"."type"='M']

25. define-mapping-class super-class:PostgradThesis class:PhDThesis map:[SELECT "Thesis"."id" FROM "Thesis" WHERE "Thesis"."type"='D']

26. add-subclass-relation super-class:User sub-class:StudentUser

27. add-subclass-relation super-class:User sub-class:TeacherUser

28. add-subclass-relation super-class:PostgradThesis sub-class:PhDThesis

29. add-subclass-relation super-class:PostgradThesis sub-class:MScThesis

30. establish-disjoint-classes class-one:MScThesis class-two:PhDThesis

31. load-excel-book from-schema:c:/magazines.xsc from-book:c:/Magazines.xlsx

32. materialize-ontology

33. add-subclass-relation super-class:Printed sub-class:Magazine

34. load-csv-file from-schema:c:/books.sch from-file:c:/books.csv

35. materialize-ontology

36. add-subclass-relation super-class:Printed sub-class:Book

37. save-ontology

Fig. 6. Script for the workflow

6 Conclusions and Future Work

We used a third-party application called OntoMetrics for conducting several experiments in the application of ontology metrics to ontologies produced with the GF system. We presented a particular case study on measuring an ontology produced in previous work, what led to adding new functionality to GF (viz., modifying domain and range of materialized properties from database relations). The results obtained show that the ontologies generated with GF are reasonably good for being used in SW applications as they are validated correctly and pass all of the filters for the main OWL2 profiles, thus making them suitable for processing with lightweight reasoners. The original experiments revealed that GF was not impacting the annotation axioms metrics, what led to further improvements allowing a semi-naive user to add annotations to ontologies, classes and properties. We also presented a scripting language that allows a semi-naive user to automate the process of OBDA that is normally performed via a GUI. Our current approach has some limitations including hardcoded names of classes and attributes, as well as limited interactions with external ontologies. Besides, as our current approach relies on materialization, it is only apt for static data sources. We are currently working on adding virtualization of external data sources for obtaining an approach more adequate to dynamic data sources. Knowledge graphs have recently obtained significant attention from both academia and industry in scenarios that require exploiting diverse, dynamic, large-scale collections of data [11]. As future work, we also plan to export some of the concepts introduced in this research line to the field of knowledge graphs.

Acknowledgments. This research is funded by Secretaría General de Ciencia y Técnica, Universidad Nacional del Sur, Argentina and by Comisión de Investigaciones Científicas de la Provincia de Buenos Aires (CIC-PBA).

References

1. Raad, J., Cruz, C.: A survey on ontology evaluation methods. In: Proceedings of the 7th International Joint Conference on Knowledge Discovery, Knowledge Engineering and Knowledge Management, November 2015
2. García, J., García-Peñalvo, F.J., Therón, R.: A survey on ontology metrics. In: Lytras, M.D., Ordonez De Pablos, P., Ziderman, A., Roulstone, A., Maurer, H., Imber, J.B. (eds.) WSKS 2010. CCIS, vol. 111, pp. 22–27. Springer, Heidelberg (2010). https://doi.org/10.1007/978-3-642-16318-0_4
3. Franco, M., Vivo, J.M., Quesada-Martínez, M., Duque-Ramos, A., Fernández-Breis, J.T.: Evaluation of ontology structural metrics based on public repository data. Briefings Bioinform. **21**(2), 473–485 (2020)
4. Bansala, R., Chawlab, S.: Evaluation metrics for computer science domain specific ontology in semantic web based IRSCSD system. Int. J. Comput. (IJC) **19**(1), 129–139 (2015)
5. Plyusnin, I., Holm, L., Törönen, P.: Novel comparison of evaluation metrics for gene ontology classifiers reveals drastic performance differences. PLOS Comput. Biol., 1–27 (2019)
6. Tovar, M., Pinto, D., Montes, A., González-Serna, G.: A metric for the evaluation of restricted domain ontologies. Comp. Sist. **22**(1) (2018)
7. Gangemi, A., Catenacci, C., Ciaramita, M., Lehmann, J.: A theoretical framework for ontology evaluation and validation. In: SWAP 2005 - Semantic Web Applications and Perspectives, Proceedings of the 2nd Italian Semantic Web Workshop, University of Trento, Trento, Italy, pp. 14–16, December 2005

8. Gómez, S.A., Fillottrani, P.R.: Specification of the schema of spreadsheets for the materialization of ontologies from integrated data sources. In: Pesado, P., Eterovic, J. (eds.) CACIC 2020. CCIS, vol. 1409, pp. 247–262. Springer, Cham (2021). https://doi.org/10.1007/978-3-030-75836-3_17

9. Gómez, S., Fillottrani, P.R.: Ontology metrics in the context of the GF framework for OBDA. In: Gaul., M.I.M. (ed.) XIII Workshop en Innovación en Sistemas de Software (WISS 2021), XXVII Congreso Argentino de Ciencias de la Computación (CACIC 2021). Red de Universidades con Carreras en Informática, Universidad Nacional de Salta, pp. 551–560, October 2021

10. Tartir, S., Arpinar, I.B., Sheth, A.P.: Ontological evaluation and validation. In: Theory and Applications of Ontology: Computer Applications, pp. 115–130 (2010)

11. Hogan, A., et al.: Knowledge graphs. ACM Comput. Surv. (2021)

Ground Segment Anomaly Detection Using Gaussian Mixture Model and Rolling Means in a Power Satellite Subsystem

Pablo Soligo$^{(\boxtimes)}$ ⓘ, Germán Merkel, and Ierache Jorge ⓘ

Universidad Nacional de La Matanza, Florencio Varela 1903 (B1754JEC), San Justo, Buenos Aires, Argentina
{psoligo,jierache}@unlam.edu.ar, gmerkel@alumno.unlam.edu.ar
http://unlam.edu.ar

Abstract. In this article we explore the possibility of finding anomalies automatically on real satellite telemetry. We compare two different machine learning techniques as an alternative to the classic limit control. We try to avoid, as much as possible, the intervention of an expert, detecting anomalies that cannot be found with classical methods or that are unknown in advance. Gaussian Mixture and Simple Rolling Means are used over a low orbit satellite power subsystem telemetry. Some telemetry values are artificially modified to generate a shutdown in a solar panel to try to achieve early detection by context or by comparison. Finally, results and conclusions are presented.

Keywords: Satellites · Ground segment · Platform · Telemetry · Machine learning · Data mining · Anomaly detection

1 Introduction

The Software for Space Research and Development Group(from Spanish Grupo de Investigación y Desarrollo de Software Aeroespacial) (GIDSA) https://gidsa. unlam.edu.ar aims to propose and test new generation space software solutions prototypes. Previous articles includes experiences and lessons learned in developing a prototype ground segment using general-purpose languages for telemetry interpretation and command script implementation, adoption of well-tested standards in the software industry, massive storage of telemetry and fault detection ([1–3]). The functioning prototype can be found in https://ugs.unlam.edu. ar which works with real satellite data, mainly obtained from SatNOGS [4].

This article is an extension of the article published in XXVII Congreso Argentino de Ciencias de la Computación 2021 (CACIC) entitled "Detección de Anomalías en Segmento Terreno Satelital Aplicando Modelo de Mezcla Gaussiana y Rolling Means al Subsistema de Potencia." [5]. As an extension of the article, external calculated data were added to the dataset and preprocessing

P. Pesado and G. Gil (Eds.): CACIC 2021, CCIS 1584, pp. 254–266, 2022.
https://doi.org/10.1007/978-3-031-05903-2_17

was applied to the original data in order to improve the detection of anomalies. The early detection of anomalies in complex system such as artificial satellites are significant. Upper and lower limits for many telemetry variables are usually the most common technique used to detect anomalous behavior [6]. As stated in a previous paper [7], the satellite health is controlled with the constant help of an expert, using little computational power. Meanwhile, in the software industry, Machine Learning is currently used for different kinds of anomaly detection, such as credit card frauds [8]. The objective of using machine learning for fault detection is mainly early detection of satellites' anomalies, avoiding, as much as possible, constant assessment from experts and also detecting types of unknown anomalies. Machine Learning offers an interesting array of possibilities in behavior prediction and anomaly detection. There are two types of machine learning algorithms: supervised and unsupervised machine learning. The former depends on labeled input data, i.e., the input dataset must have defined whether a datum is considered an anomaly or not. The latter does not depend on labeled input data, instead, it learns the internal representation of the dataset and generates patterns [9]. An anomaly is any datum that deviates from what it is expected or normal. In the statistics literature, they are also referred to as outliers. Every datum that is processed by the prototype will be classified using binary labels: a datum is an anomaly or not [10]. For detecting anomalies, fault detection machine learning algorithms create a model of the nominal pattern in the dataset, and then compute an outlier score of a given datum. Depending the algorithm, this outlier score takes into account the correlation with different features or not [10]. In this work and in the case of time series data, we look for a sequence of outliers that determines an anomaly rather than a particular piece of data; we look for abnormal system behavior instead of a wrong value. In previous articles we proposed guidelines and developed a prototype of a multi-mission ground segment [2] which included limit control for the telemetry variables. Then we set dynamic limits using machine learning [7]. The current work explores a prototype for telemetry mining and health control for a multi-mission ground segment using two different machine learning algorithms: Gaussian Mixture and Rolling Means. These two algorithms are investigated and compared one another to study the feasibility of applying them in a real-time health control, and pattern and behavior detection prototype.

In Sect. 2 we describe the dataset we use for telemetry mining, the two main algorithms, and three other algorithms that were discarded. We extended this section including external data with information about satellite's eclipse. We also added a third algorithm to Sect. 2.3, Density-Based Spatial Clustering of Applications with Noise (DBSCAN), and Principal Components Analysis (PCA) for preprocessing the dataset before applying Gaussian Mixture. Section 3 shows the results of applying Gaussian Mixture and Rolling Means to the dataset for telemetry mining. We extended this section including with the results of applying PCA preprocessing before Gaussian Mixture. In Sect. 4 we describe conclusions obtained from the results presented for Rolling Means and Gaussian Mixture. In this section, we added conclusions from using PCA preprocessing before applying

Gaussian Mixture. Finally, 5 presents ways to improve the current work results using automation for detecting number of components to be used when applying PCA.

2   Materials and Methods

For this experiments we use our own dataset with real telemetry data https:// gidsa.unlam.edu.ar/data/LowOrbitSatellite.csv. The source of telemetry is a scientific low orbit satellite. Telemetry begins at 2015-05-27 08:51:06+00:00 and ends at 2015-06-05 23:34:06+00:00. For these experiments we use only the first two days, from 2015-05-27 08:51:22+00:00 to 2015-05-29 08:50:59+00:00. The training dataset finally has 17277 rows, the test dataset has 4320 rows. Table 1 shows the feature meanings, all values, except *vBatAverage* and *BatteryDischarging* are in raw, however data is always normalized before processing.

Unfortunately, the dataset does not have any expert-documented or probable failures. As an extension external data has been incorporated, using python PyEphem [11] library, we added two features *cInEclipse* and *elapsedTime*.

cInEclipse indicates if the satellite is in eclipse or not, and *elapsedTime* indicates how long the satellite is eclipsed (negative values in seconds) or not (positive values in seconds). Dataset with additional features is available at https://gidsa.unlam.edu.ar/data/LowOrbitSatelliteWithEclipses.zip. Due to the fact that there are no probable or documented failures in the dataset, we partially cut off the power supply from solar panel 24 turning 0 the current (128 in raw) to create an artificial anomaly in test dataset. The cut off is progressive and covers 1079 rows. This is similar to leaving the panel eclipsed, regardless of the real context. Note that classical limit control cannot handle this behavior, as currents close to 0 are perfectly valid on eclipse periods.

Table 1. DataSet features

Feature	Meaning
vBatAverage	Average of Battery voltage used by supervisions
BatteryDischarging	Flag True/False if battery is discharging
ISenseRS1	IsenseRS1 current (battery current)
ISenseRS2	IsenseRS2 current (battery current)
V_MODULE_N_SA0 < N < 25	Current in solar panel #N

All telemetries from the Power Control System (PCS/EPS) are highly correlated as shown in Fig. 1, for size reasons we show correlation from only 4 of the 24 panel current features.

The current work shows two different machine learning algorithms to detect anomalies in satellite telemetry: Gaussian Mixture and Rolling Means. The former is applied to the telemetry system of a satellite as a whole, using the correlation between variables, while the latter is applied to every telemetry variable.

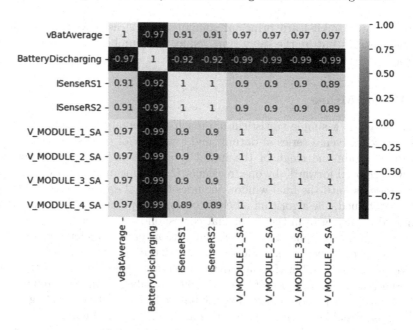

Fig. 1. Correlation between PCS features

Both of these approaches follow classic statistical approaches: both use classic statistical measures like mean, standard deviation and probability.

Finally, and also as an addition to the original article, scikit-learn [12] principal components are applied to the dataset in order to reduce the dimensions and achieve better results.

2.1 Gaussian Mixture Model

Using scikit-learn [12] library, we create a Gaussian Mixture Model (GMM). GMM can be used to cluster unlabeled data, GMM can help to detect behavior that is far or unlikely to nominal behavior. Any point which is very far from the established clusters could be considered an anomaly.

The dataset was divided into two sub-datasets: training dataset and test dataset. The test is made over two days of telemetry. Dataset's last 20% forms the test dataset, while the other 80% forms the training dataset. A model is obtained by running the algorithm over the training set, where the components amount and the covariance type are selected in an iterative process that analyzes the information-theoretic criteria (BIC), covering the 4 covariance types and the components amount from 1 to 20. The minimum outlier score is set as the limit for the test dataset. As an extension of the original article we added two external-calculated features. By adding two more features (*cInEclipse* and *elapsedTime*), information typically available in the flight dynamics departments we introduce correlated external information to help the algorithm. In addition, we incorporated scikit-learn principal components [12] as a way to utilize all available features but at the same time reduce the number of dimensions and help the algorithm.

2.2 Rolling Means

Rolling Means is a simple statistical approach to anomaly detection in a time series dataset. Given a series of datums and a window of fixed size N, the algorithm first obtain the mean of the initial N datums in the series. Then the window is "moved forward" by one, recalculating the mean of the window. This process repeats until the final window includes the final datum. Once all means has been obtained, the algorithm labels as outliers all points that are more than S standard deviation from the mean that corresponds to the point.

Rolling Means is applied to each sub-dataset, one for each telemetry variable, and for each one it generates a model of normal. To use this algorithm, a fixed window size and a fixed number of standard deviations must be set. To decide the value of these parameters, various iterations with different values on the same dataset are run, and finally, with human inference, the best values are set. The rolling means algorithm is available at https://gidsa.unlam.edu.ar/data/rolling.py.

Rolling Means was chosen since it is a sensitive-to-outliers algorithms, while still being simple. It relies on standard deviation, taking into account the change in the time series using the fixed-size window.

2.3 Other Methods

Multivariate Normal Distribution method was also tested, but it was discarded in favour of Gaussian Mixture, due to the fact that the former needs its data to follow a normal distribution, and cannot handle multiple bells. Isolation Trees was also tried, but was discarded early given that it mislabeled most of the "normal" dataset as anomalies.

Finally, we also considered DBSCAN, another popular clustering algorithm that uses density of datums to detect anomalies. These clusters can be of any shape and are detected automatically by the algorithm, given a constant that determines the distance between points to be considered in the same cluster. The implementation of this algorithm has two problems: it does not handle high dimensionality well, and it mislabels "normal" datums as anomalies because they are not found in dense clusters. The second problem can be seen in Fig. 2, where datums that represent the transition in panel 24 current are mislabeled as anomalies. Taking into account this problems means using a larger constant for the distance between points, but it results in non-detection of artificial anomalies.

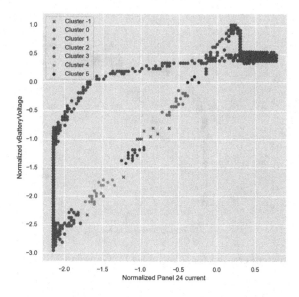

Fig. 2. Clusters created for 2 variables by DBScan

3   Results

3.1   Gaussian Mixture Model

For visualisation purposes the final model with training data and only two features, *V_MODULE_24_SA* and *vBatAverage*, are shown in Fig. 3. Gaussian functions are shown in green and they include the clusters.

Figure 4 shows the results of applying the model with 2 specially selected features to the test data. The dataset was not artificially changed, it fits perfectly to the created model.

Results with modified dataset are shown in Fig. 5. 880 anomalies are detected and flagged with a cross-mark. Since the current drop is progressive, it cannot be precisely established which of all the data are anomalies and which ones are not. On the other hand, in previous articles [5], it was established that using one set of characteristics produced 11 false positives on the test data of the original dataset, about 1% from modified rows.

Using the dataset with external and internal features, without principal components and without anomalies the algorithm generates 8 and 3 false positives as shown in Table 2. Information about amount of mixture components (GM Comp) and type of covariance (Cov.Type), automatically obtained, is also added to the table.

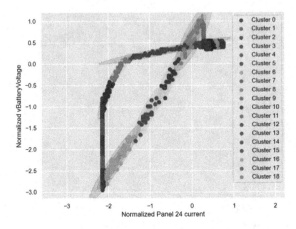

Fig. 3. Clusters created for 2 variables by Gaussian Mixture Model

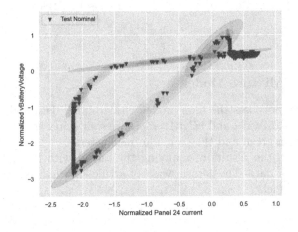

Fig. 4. Test dataset without anomalies

Table 2. Anomalies detected with GMM using 2 (*V_MODULE_24_SA* and *vBatAverage*), 36 features (Internal), 38 features (Internal and external) and original dataset

#Feat.	Anom.Train	Anom.Test	GM Comp	Cov.Type
2	0	0	14	Diag
36	0	8	18	Full
38	0	3	16	Full

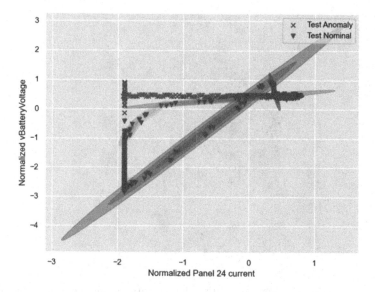

Fig. 5. Test dataset with anomalies

The same experiment on artificially modified data detects 915 and 916 anomalies out of a total of 1079 modified records. Not all modified records are in error condition respect to the system because the cut is progressive. Table 3 shows the results.

Table 3. Anomalies detected with GMM using 2 (*V_MODULE_24_SA* and *vBatAverage*), 36 features (Internal), 38 features (Internal and external) and artificially modified dataset

#Feat.	Anom.Train	Anom.Test	GM Comp	Cov.Type
2	0	849	19	Diag
36	0	915	18	Full
38	0	916	19	Full

Using the dataset with external and internal features, principal components, without anomalies the algorithm generates 3 false positives as shown in Fig. 6. Note that to have a drawable result the number of components used in principal components is equal to 3.

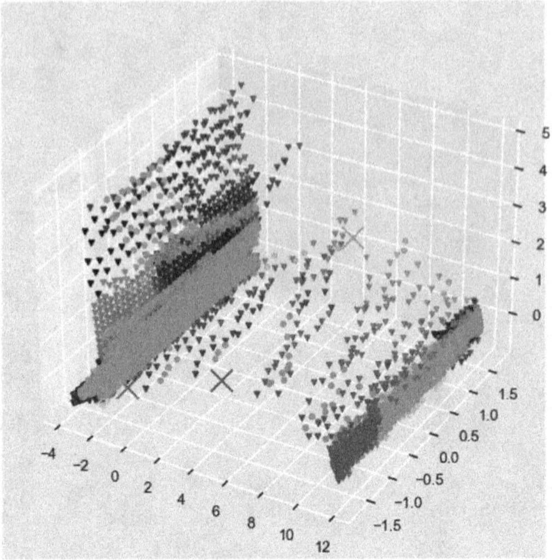

Fig. 6. Gaussian Mixture clusters with original dataset and principal components (n = 3). The x-marks represent false positives

Finally, Tables 4 and 5 show the results of the experiment using principal components with number of components between 3 and 6 for internal characteristics and for internal and external characteristics.

Table 4. Anomalies detected with GMM and principal components using 36 features (Internal), 38 (Internal and external) features and the original dataset

#Feat.	#Components	Anom.Train	Anom.Test	GM Comp	Cov.Type
36	3	0	3	19	Full
36	4	0	1	19	Full
36	5	0	0	19	Full
36	6	0	0	19	Full
38	3	0	1	19	Full
38	4	0	0	19	Full
38	5	0	3	17	Full
38	6	0	4	19	Full

Table 5. Anomalies detected with GMM and Principal components using 36 features (Internal), 38 (Internal and external) features and the artificially modified dataset

#Feat.	#Components	Train	Test	GM Comp	Cov.Type
36	3	0	2	18	Full
36	4	0	920	19	Full
36	5	0	922	18	Full
36	6	0	923	19	Full
38	3	0	0	19	Full
38	4	0	989	19	Full
38	5	0	927	19	Full
38	6	0	928	18	Full

3.2   Rolling Means

The dataset is divided into 7 sub-datasets, one for each telemetry variable. For each of these sub-datasets, the algorithm Rolling Means labels the datums of each variable as anomalies or not, based on the "model of normal".

Table 6. Anomalies detected with Rolling Means

# of Standard deviations	Normal dataset	Anomaly dataset
1	260	449
2	71	6

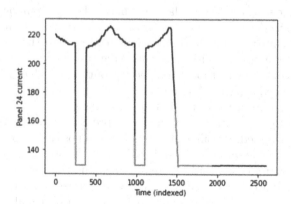

Fig. 7. Rolling Means applied to Panel 24 using 1 number of standard deviations

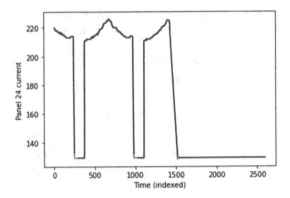

Fig. 8. Rolling Means applied to Panel 24 using 2 number of standard deviations

Running the algorithm, with a window size of 1000 (a window that is half the number of inserted anomalies), using 1 and 2 number of standard deviations lead to the next results. Each sub-dataset is plotted with blue lines, and the red dots are the anomalies that the algorithm detected. The results are showed in Table 6 (Figs.7 and 8).

4   Conclusions

In the case of the Rolling Means, the model doesn't have sense of correlation, and labels as anomaly based on the trend of the time series dataset. An isolated spike in the graph will be labeled as an anomaly, but may be the result of a contextual expected action. Even though Rolling Means is a simple algorithm, with low computational needs, not taking into account the context and correlations cannot handle fine-grained, context-dependant anomalies. Using 1 number of standard deviation seems to detect the introduced anomalies, but it also mislabels valid data. Using 2 number of standard deviations, contrary to the expected, turns out to behave the same way, mislabeling more datums than detecting anomalies. Rolling Means is a valid method for detecting outliers produced by noise, but cannot be considered a valid algorithm for detecting anomalies. It also needs the intervention of an expert to set the initial parameters.

On the other hand, Gaussian Mixture shows promising results, the introduced anomalies were detected, and in some cases no false positives were found. Using the dataset, without external data and reducing with principal components to 5 or 6 dimensions, no false positives are obtained and the vast majority of artificially modified data is detected as anomalous (Tables 4 and 5). In addition, using the dataset with external data and reducing with principal components to 4 dimensions, no false positives are obtained and also the vast majority of artificially modified data is detected as anomalous (Tables 4 and 5). It is important to note that the introduced anomalies are not detectable by a limit control system because the values may be normal in certain contexts. The covariance and amount of clusters were obtained automatically, without an expert intervention. The results give an informative view of the different algorithms, but cannot be objectively evaluated since there are no labeled data available to compare them to. If labelled data were available, a more accurate conclusion could be reached. The use of principal components allowed for no false positives in the original dataset.

5   Future Works

The current work presents a prototype of a satellite anomaly detection system, which is part of a next generation ground segment [2]. By no means this is an exhaustive exploration of what machine learning can do in the field of anomaly detection, but may represent a feasible alternative to limit control that is in use today. Also, as previously stated [7] the prototype doesn't need help of from an expert in an operation setting, but needs an initial help on setting the parameters for both models and analyzing its results.

The biggest barriers found are the advanced machine learning expertise to optimize the algorithms and the lack of pre-labeled data. More complex models could be explored, namely AutoRegressive Integrated Moving Average (ARIMA), or Neural Network models. In the case of Gaussian mixture the results are promising, but the process of selecting the number of components in principal components should be automated in the same way as selecting the number of components in Gaussian Mixture. Also, it is required to test with new real and ideally labelled datasets as well as different splits of the datasets to avoid overfitting.

References

1. Soligo, P., Ierache, J.S.: Software de segmento terreno de próxima generación. In: XXIV Congreso Argentino de Ciencias de la Computación (La Plata, 2018) (2018)
2. Soligo, P., Ierache, J.S.: Segmento terreno para misiones espaciales de próxima generación. In: WICC 2019 (2019)
3. Soligo, P., Ierache, J.S., Merkel, G.: Telemetría de altas prestaciones sobre base de datos de serie de tiempos (2020)
4. Satnogs satnogs. https://satnogs.org/. Accessed 30 July 2021
5. Soligo, P., Merkel, G., Ierache, J.: Detección de anomalías en segmento terreno satelital aplicando modelo de mezcla gaussiana y rolling means al subsistema de potencia. In: XXVII Congreso Argentino de Ciencias de la Computación (CACIC) (Modalidad virtual, 4 al 8 de octubre de 2021) (2021)
6. Yairi, T., Nakatsugawa, M., Hori, K., Nakasuka, S., Machida, K., Ishihama, N.: Adaptive limit checking for spacecraft telemetry data using regression tree learning. In: 2004 IEEE International Conference on Systems, Man and Cybernetics (IEEE Cat. No. 04CH37583), vol. 6, pp. 5130–5135. IEEE (2004)
7. Soligo, P., Ierache, J.S.: Arquitectura de segmento terreno satelital adaptada para el control de límites de telemetría dinámicos (2019)
8. Rosenbaum, A.: Detecting credit card fraud with machine learning (2019)

9. Hastie, T., Tibshirani, R., Friedman, J.: The Elements of Statistical Learning. Springer Series in Statistics, 2 edn. Springer, New York (2008). https://doi.org/10.1007/978-0-387-84858-7

10. Aggarwal, C.C.: An introduction to outlier analysis. In: Outlier Analysis, pp. 1–34. Springer, Cham (2017). https://doi.org/10.1007/978-3-319-47578-3_1

11. Rhodes, B.C.: Pyephem: astronomical ephemeris for python. Astrophysics Source Code Library, pp. ascl-1112 (2011)

12. Pedregosa, F., et al.: Scikit-learn: machine learning in Python. J. Mach. Learn. Res. **12**, 2825–2830 (2011)

Signal Processing and Real-Time Systems

Prototype of a Biomechanical MIDI Controller for Use in Virtual Synthesizers

Fernando Andrés Ares[1], Matías Presso[1,2,3(✉)], and Claudio Aciti[1,2]

[1] Universidad Nacional de Tres de Febrero, Sede Caseros I, Valentín Gómez 4828, B1678ABJ Caseros, Buenos Aires, Argentina
ferares4@gmail.com, matiaspresso@gmail.com,
caciti@exa.unicen.edu.ar
[2] Universidad Nacional del Centro de La Provincia de Buenos Aires, Pinto 399, 7000 Tandil, Buenos Aires, Argentina
[3] Comisión de Investigaciones Científicas de La Provincia de Buenos Aires, Calle 526 entre 10 y 11, 1900 La Plata, Buenos Aires, Argentina

Abstract. This paper describes the design, development, and implementation of a prototype for a low-cost, biomechanical device. Using the MIDI protocol, this device is capable of controlling, through body movements, different types of synthesizers or virtual instruments currently available in the music industry. Some aspects are discussed in connection with the design, the choice of components, the manufacture of a printed circuit board, and an ergonomic mounting case designed for this purpose. Finally, metrics and verifications of the prototype's operation were carried out, by integrating communication and audio software tools.

Keywords: MIDI controller · Biomechanics · Body movement · Synthesizer · Virtual musical instruments

1 Introduction

Since the 1980s, all kinds of tools and controllers have been manufactured using the Musical Instrument Digital Interface (MIDI) [1] protocol for various functionalities. The MIDI protocol is a music industry standard used to connect digital musical instruments, computers, and various mobile devices. The vast majority of the tools are developed in the format of pianos or other types of instruments with keys, buttons, sliders, or other mechanisms. However, devices having these characteristics but activated exclusively with body movements are not abundant in the current market.

The inception of the MIDI protocol, as we know it today, is attributed to Dave Smith and Chet Wood [2] in collaboration with different companies operating in the market in the 1980, including Roland, Yamaha, Korg, Kawai, among others. This interface was a solution to a common problem in response to the rise of analog synthesizers. The protocol was intended to generate a common language for the interconnection of this type of instrument, which until then, had followed different standards according to the manufacturer. The first version of the MIDI protocol was announced to the public in

1982, and it was only at the end of that year that the first instrument implementing such a protocol was launched in the market.

Since then, the protocol has evolved either with improvements in its specification or new functionalities adapting to contemporary technologies. Today, this protocol is used not only for the interconnection of devices but also as a common language for digitally creating, performing, and broadcasting music. Some companies like Holonic Systems [3] and projects like "Making music with your muscles!" [4], "A MIDI Controller based on Human Motion Capture" [5], "Wireless Midi Controller Glove" [6], "Glove MIDI Controller" [7], and "KAiKU Glove Wearable MIDI Controller - S" [8] address the same subject but have perspectives and purposes that differ from those of the present work. The present paper is an extension of our previous work published in CACIC2021 [9] and includes a public access repository and new features such as new diagrams, 3D designs, functional verification of the device and latency metrics.

2 Motivation, Objectives, and Scope

It is currently difficult to find a MIDI controller that works exclusively with body movements, and existing products do not easily reach consumers due to cost and flexibility issues or acquisition difficulties. The motivation of this study was to build a low-cost device with easily accessible materials or components. The goal was to develop a device for generating new forms of musical creation and performance through the use of body mechanics. In this way, the device is aimed at enriching the spectrum of current music-making practice through a paradigm shift in how to play a virtual instrument. To this end, the specific objectives were as follows:

1. To design a hardware device that has a three-dimensional motion sensor.
2. To develop a mechanism to translate the movements captured by the hardware device to the MIDI protocol.
3. To create a mechanism for transmitting the MIDI signal to a PC through a wireless medium.
4. To design a case for strapping the device to the back of the hand.
5. To verify and test the functioning of the device using support software that simplifies the wireless connection between the device and a PC, including metrics to analyze the latency and response times of the design during its operation.

The scope of the work is limited to the design of the device and the development of basic functionalities with certain defined parameters that can be tested in a prototype. It does not include the development of a specific driver for handling the hardware from a PC or the handling of virtual ports for use from any MIDI application. The prototype was evaluated in a work environment with a PC and Windows 10 operating system, high signal intensity, and sensors within an interference-free environment.

3 MIDI

MIDI is a standard protocol for serial communication in the music industry designed for the interconnection of musical instruments and devices. The information is transmitted in the form of binary-coded messages, which can be considered instructions that tell a device how to play a piece of music with certain parameters defined by the protocol.

MIDI devices can transmit, receive, and retransmit messages according to the nature and function of the particular device. Some devices are designed solely to create and transmit MIDI messages and can be connected to a slave device, capable of reproducing sound upon receiving these messages. This type of device, which issues orders for the execution of instructions, is known as MIDI controller (Fig. 1).

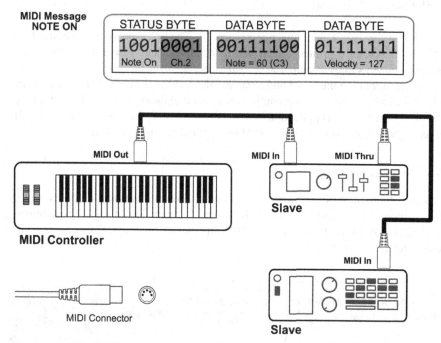

Fig. 1. Connections, connector, controller, and MIDI messages.

4 Problem Description

The problem addressed in this work involves building a device capable of translating the movements of a hand into sounds, through a virtual synthesizer installed on a computer with a Microsoft Windows operating system. It must be built using low-cost materials available in the market. It must also be able to connect wirelessly to a computer to transmit the data associated with each movement. In addition, communication must be conducted by means of the MIDI standard so that it can be supported by any virtual synthesizer currently available.

Three aspects must be considered in the design and construction of the device. First, a hardware solution must be designed and tested to interconnect all the components required for the device to work. Second, a support case must be designed so that it can be strapped to one hand. Third, a software solution capable of translating hand movements into MIDI-type messages must be developed.

The software should include a calibration mechanism to ensure that the data obtained are accurate and that MIDI messages are sent at valid times in a real-time context for audio applications. Being a wireless device, it must be powered by batteries allowing reasonable autonomy and the possibility of opting for rechargeable batteries.

5   Proposed Solution

The solution consists of three fundamental aspects. The first aspect comprises the design and construction of a printed circuit board for all the hardware components and its respective power supply circuit. The second aspect involves the design and manufacture of a case for screwing the board and strapping the hand to the case. This case must also contain the battery and any other component integrated into the device. The third aspect is the development of an application that translates the movement of the hand into MIDI messages, which will then be interpreted by a virtual instrument running on a PC.

5.1   System Components

The block diagram in Fig. 2 shows an overview of how the various components of the device are interconnected, as well as the external connection to a PC. In addition, it shows the hardware components of the prototype device and the software tools used in the computer.

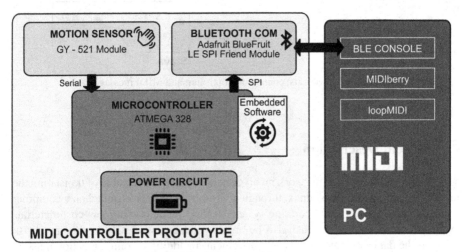

Fig. 2. Block diagram of the system.

As for the hardware of the controller device prototype, the functional design is divided into four main blocks, each for a specific purpose: motion detection, power management, communication, and processing.

To capture the three-dimensional movements of a hand, it is essential to have a sensor capable of generating a real-time representation of the variations in the position or rotation of the hand. A low-cost sensor that is reliable enough to avoid noise in the measurement must be chosen among the options currently available. The MPU-6050 sensor [10] of InvenSense combines the features mentioned. It is composed of a three-axis accelerometer that senses gravitational acceleration, a three-axis gyroscope that measures rotational speed, and an integrated Digital Motion Processor (DMP), which simplifies the management of calculations to combine the measured data in real-time. Among the most conventional modules, the one chosen for the prototype was the GY-521 [11].

A fundamental aspect of the controller is the communication between the device and a computer. Wireless communication is ideal because it provides greater comfort and portability, and Bluetooth technology is the global standard that offers such features. One of the most recent improvements to this technology is called Bluetooth Low Energy or BLE. Several types of wireless communication modules use this technology. Different models were evaluated for this project and the chosen one was the Adafruit Bluefruit SPI module [12], which establishes synchronous and serial communication for data transmission. It has substantial advantages because the firmware is fully developed for this module, which makes it reliable, robust, and easy to configure and allows communication through the SPI protocol.

For the processing and execution of the device program, the ATMEGA328P [13] microcontroller was chosen because it is a low-power, low-cost, and perfectly accessible chip for this project. It has several inputs and outputs compatible with the sensors, which enables the future expansion of the device. It also has advanced programming tools.

To supply energy, a power-supply circuit was designed using as small an external battery as possible, with reasonable autonomy. Consumption measurements were made based on the normal use of the device, which showed current oscillations in a range from 49 to 51 mA. For a conventional 9-V battery and an approximate capacity of 500 mAh, autonomy was estimated to be about 10 h.

The connection between the device and a PC requires support software. Three software tools were used for different purposes: BLE Console [14] was used to connect the Bluetooth Adafruit BlueFruit LE SPI Friend module to the operating system; loopMIDI [15] was used to create and emulate a virtual MIDI port; and MIDIBerry [16] was utilized to map inputs and outputs from the module to the virtual port.

5.2 Motion Capture

In order to sense the movements of a hand in three dimensions, the yaw, pitch, roll, or YPR methodology, was used, as shown in Fig. 3.

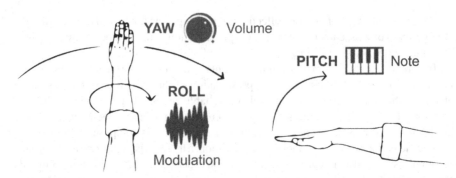

Fig. 3. YPR representation [17] of the hand movements and associated parameters of the virtual MIDI controller.

Given that the movements of the human arm are restricted, the measurement radius of each of the axes must be limited to a range of 0 to 180°. Each of the variations with respect to the axes is associated with a certain event, which is translated into a MIDI message. Three variable parameters are proposed for this work. The parameter of the notes is associated with the vertical movement of the hand, defined by the pitch variable. In this case, the range of notes of a Grand Piano is between the values 21 and 108, so a linear transformation is carried out to dynamically adjust values from 0 to 180 to the previously mentioned note range. According to the MIDI specification, this parameter is transmitted using a Note on/off message with the respective note value.

For volume, the MIDI specification requires a dynamic range represented by values between 0 and 127, and a similar linear transformation is performed. This parameter is associated with the horizontal movement of the arm, or yaw. This message is transmitted via the MIDI velocity parameter, and it is added as part of the Note on/off message data.

The modulation of the note is associated with the rotation of the hand. The MIDI parameter to be used is specified by means of a Control Change that uses the range from 0 to 127. A linear transformation is also carried out.

6   Software Controller

The MIDI controller prototype software comprises three stages: an initial configuration and initialization of the microcontroller and sensor variables; a second calibration of the device; a main loop to sense the movements and send messages wirelessly.

The configuration stage involves the preliminary settings for program execution, such as the initialization of the serial port, for the correct functioning of the application. The Wire library is initialized to capture the sensor data and the first calibration of the device is performed. Finally, the Bluetooth module is initialized and the connection is verified before starting the transmission.

Calibration is performed with the device at rest during initialization. A series of parameterizable samples are considered and averaged, and then the values obtained are subtracted from each measurement in real-time.

The main loop of the program consists of the following sequence: reading of accelerometer and gyroscope records, calculation of angles in each axis, complementary filtering, linear transformation for value adjustments, and transmission of the MIDI message.

Reading of Accelerometer and Gyroscope Records: The motion sensor records are accessed directly through the Wire library [18]. In this way, any external factor that may affect the total time interval of the application is avoided. The records are read sequentially in each program cycle.

Angle Calculations: According to the sensor information and using the factory settings, it is possible to obtain the acceleration variation in each axis by dividing the value measured by the sensitivity unit.

Complementary Filtering: Both accelerometer and gyroscope data are prone to systematic errors. To solve this problem, a standard method is used, whereby both measurements are combined by using complementary filters to obtain a single value.

Linear Transformation: The scale adjustment on each axis, taking angles from 0° to 180° for each MIDI parameter, is performed with a linear transformation to compress or expand such values according to the new scale, using the map function () from the Math library [19].

```
int note = map(pitch,-90,90,21,108);
int velocity = map(roll,-90,90,127,0);
int modulation = map(yaw,-90,90,0,127);
```

Transmission Of MIDI Messages: A library supplied by the Bluefruit module Adafruit_BLEMIDI [20] is used to instantiate the MIDI service and send messages. These messages are sent by invoking one of its methods:

```
// note on
midi.send(0x90, note, velocity);

// note off
midi.send(0x80, note, velocity);

// control change
midi.send(0xB, controller_number, ccvalue);
```

7 Final Prototype

To build the final prototype, a three-dimensional model and a preliminary prototype were previously developed, and their functional and ergonomic aspects were evaluated (Figs. 4, 5, 6, 7, 8 and 9).

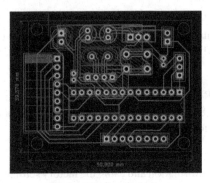

Fig. 4. PCB board design

Fig. 5. 3D circuit design

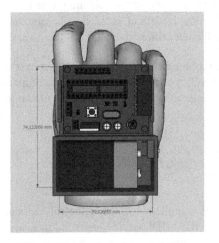

Fig. 6. 3D top view of the simulation

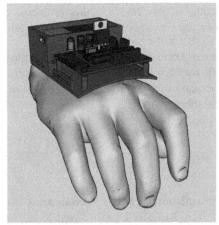

Fig. 7. 3D perspective view

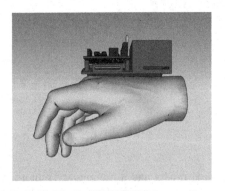

Fig. 8. 3D left side view

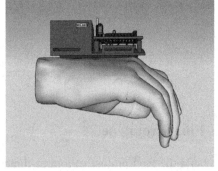

Fig. 9. 3D right side view

The final prototype was built using a simple-face, FR4 fiberglass printed circuit board, with a 50.8 mm by 39.4 mm solder mask, reducing the overall size of the device. Finally, the board was screwed to a three-dimensional printed model and mounted on the hand using two Velcro straps (Figs. 10–11).

Fig. 10. Aerial view of the device

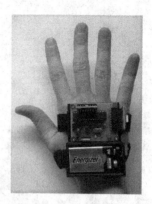

Fig. 11. Device strapped to the hand

8 Functional Verification of the Device

To verify the operation, we used the support software tools mentioned above. BLE Console was used to establish communication with the controller device, loopMIDI to emulate the virtual MIDI port, and MIDIBerry to map the input/output of the virtual port (Figs. 12, 13 and 14). There are alternative support tools, such as the Hairless MIDI Serial Bridge [21], a multiplatform alternative to the MIDIBerry option that binds ports in Microsoft Windows. In addition to the support applications for evaluating the prototype and generating sounds, a virtual instrument must be used, which can be included in any digital audio workstation software (Digital Audio Workstation, DAW) with Virtual Studio Technology (VST) support. We used virtual instruments and sounds within the free version of the SampleTank Custom Shop audio application.

Once the prototype was built, the latency of the device was estimated and analyzed. After the prototype was calibrated, the peak and average latency values were acquired as parameters for a certain number of samples. For this purpose, we introduced a modification to the testing software, by setting a mark before the beginning of the register reading and after the delivery of any type of MIDI message. These two marks are measured in milliseconds, and the peak latency is calculated by the subtraction between the finish time and the start time of each cycle. In parallel, a second parameter is defined to average these measurements by adding the times corresponding to each sample to the previous result and dividing it by the sum of the accumulated samples. After several measurements of the delay for different pitch, roll, and yaw movements, values were observed in the range of 2–9 ms, with an average value of 4 ms. Figure 15 shows the delays for different samples.

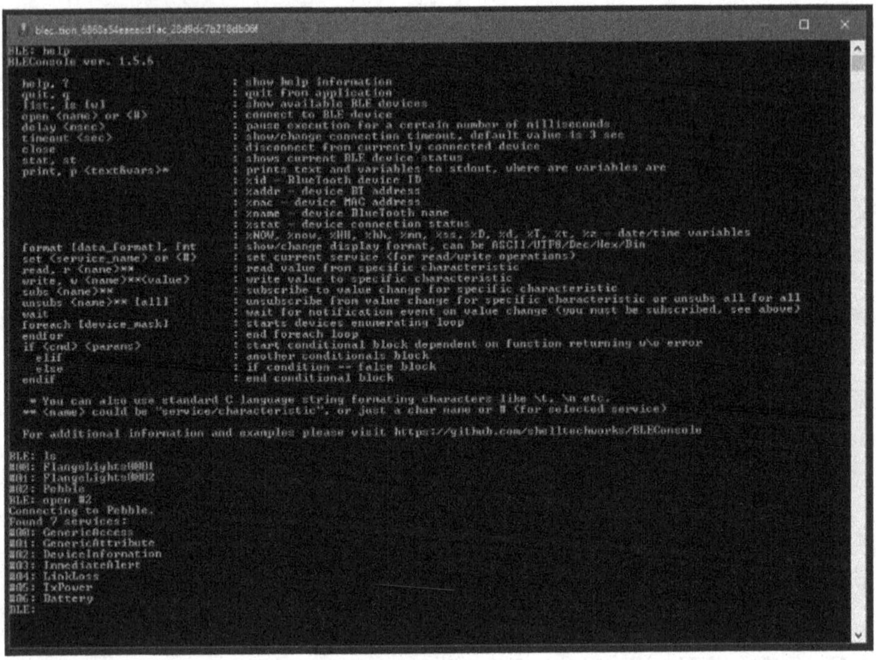

Fig. 12. BLE console user interface

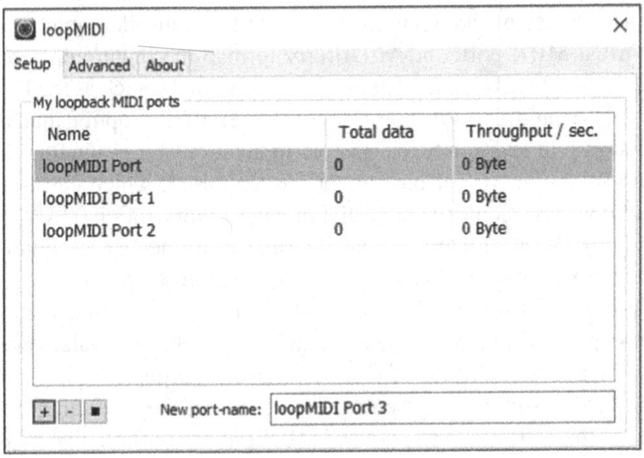

Fig. 13. LoopMIDI user interface

The measured latency values can be considered reasonable and appropriate for the device because the time delays that can be perceived by the human auditory system start at values of 12–15 ms, and they depend, in turn, on the hearing acuity of the listener.

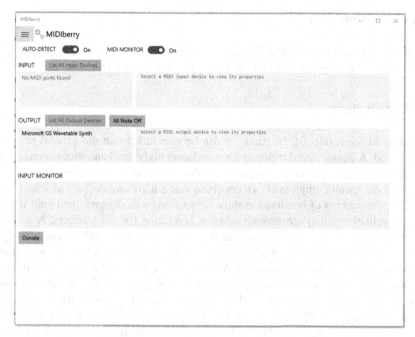

Fig. 14. MIDIberry user interface

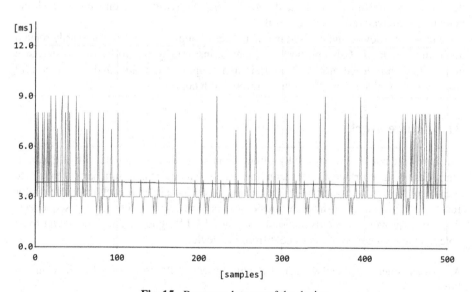

Fig. 15. Response latency of the device

9 Public Repository

All information and data is available in a public github repository of this project [22]. Firmware code, demos, case and board design are included for public access.

10 Conclusions

According to the results of this study, it can be concluded that the general objective was fulfilled. A device was developed for translating body mechanics into sound, using different types of standard tools that are available in the current music industry.

As for the specific objectives, we conclude that a hardware device capable of capturing the movements of one hand in three dimensions was designed and built using a low-cost, reliable motion sensor with inherent correction for noisy effects. A software solution was also developed to transform the data obtained into MIDI-type messages. A wireless transmission mechanism using low-power Bluetooth technology was implemented to optimize resource utilization. Additionally, a case was built using 3D printing technology for an ergonomically designed hand-strapping device. This device was tested in different iterations, adding value to the design process and improving the final product. It is worth noting that all the stages of the entire manufacturing process were experienced, generating a tangible and suitable result for evaluating the product, both for the academic and commercial fields.

In addition, the final prototype was entirely built using inexpensive and easily accessible components and materials. Therefore, its manufacture can be carried out at a low cost and in a short period, with the advantage offered by the experimentation with new technologies, such as the design and manufacture of three-dimensional models, adding a focus on innovation to the final result.

Finally, the device can be used in other types of applications, not necessarily related to the music industry. Other purposes that may be inspired by this work include assistance to people with reduced mobility, entertainment, augmented reality, and remote control applications, which are useful in the current health context.

11 Future Works

The hardware could be improved by adding manual controls to regulate software parameters, designing a surface mount technology board, and using a lithium battery to reduce the size of the device. As for the software, it would be possible to develop an arpeggiator functionality to generate different music note patterns, such as chords and sequences. Another improvement entails incorporating the MIDI Program Change parameter to be able to change the type of instrument from the device.

Acknowledgments. The authors thank CIC PBA, where M. Presso works as Research Support Professional.

References

1. The Complete MIDI 1.0 Detailed Specification: The MIDI Manufacturers Association (1996)
2. Smith, D., Wood, C.: The 'USI', or Universal Synthesizer Interface, AES (1981)
3. Holonic Systems. https://www.holonic.systems/
4. ARDUINO TEAM (2018). https://blog.arduino.cc/2018/06/04/making-music-with-your-muscles/. Arduino - https://www.arduino.cc/reference/en/. MIDI - https://ccrma.stanford.edu/~craig/articles/linuxmidi/misc/essenmidi.html
5. Velte, M.: A MIDI controller based on human motion capture (2012). https://doi.org/10.13140/2.1.4438.3366
6. Brady, M., Palecki, S., Belfort, A.: Wireless Midi Controller Globe (2018). https://courses.engr.illinois.edu/ece445/getfile.asp?id=12401
7. Dorsey, A., Gunther, E., Smythe, J.: Glove MIDI Controller. https://people.ece.cornell.edu/land/courses/ece4760/FinalProjects/s2010/ecg35_ajd53_jps93/ecg35_ajd53_jps93/index.html
8. The Music Glove, Kaiku. https://www.kaikumusicglove.com/
9. Ares, F.A., Presso, M., Aciti, C.: Prototipo de controlador MIDI biomecánico para uso en sintetizadores virtuales. In: Acta de memorias del XXVII Congreso Argentino de Ciencias de la Computación, pp. 633–642 (2021). ISBN 978-987-633-574-4
10. MPU6050 - Register Map and Descriptions Revision 2.1. https://invensense.tdk.com/products/motion-tracking/6-axis/mpu-6050/. https://invensense.tdk.com/wp-content/uploads/2015/02/MPU-6500-Register-Map2.pdf. https://playground.arduino.cc/Main/MPU-6050/
11. Tutorial: How to use the GY-521 module (MPU-6050 breakout board) with the Arduino Uno, Michael Schoeffler, Consultado Enero (2020). https://www.mschoeffler.de/2017/10/05/tutorial-how-to-use-the-gy-521-module-mpu-6050-breakout-board-with-the-arduino-uno/
12. Townsend, K.: Introducing the Adafruit Bluefruit LE SPI Friend (2020). https://cdn-learn.adafruit.com/downloads/pdf/introducing-the-adafruit-bluefruit-spi-breakout.pdf?timestamp=1594981255
13. ATMEGA328. https://www.microchip.com/wwwproducts/en/ATmega328P. https://drive.google.com/file/d/1ydLJbmDPw5O1KUB8RgsAxUxedzD1G0qI/view
14. Ble Console. https://sensboston.github.io/BLEConsole/
15. virtualMIDI, Tobias Erichsen. http://www.tobias-erichsen.de/software/virtualmidi.html
16. How to use MIDIberry with DAW. http://newbodyfresher.linclip.com/how-to-use-with-daw
17. GUI without the G: Going Beyond the Screen with the Myo™ Armband. https://developerblog.myo.com/gui-without-g-going-beyond-screen-myotm-armband/
18. Wire Library. https://www.arduino.cc/en/reference/wire
19. map(), Arduino. https://www.arduino.cc/reference/en/language/functions/math/map/
20. Adafruit BLE Library Documentation, Dan Halbert, June 2020. https://readthedocs.org/projects/adafruit-circuitpython-ble/downloads/pdf/latest/
21. The Hairless Midi to Serial Bridge. https://projectgus.github.io/hairless-midiserial/
22. Biomidicontroller repository. https://github.com/biomidiproject/biomidicontroller

Computer Security

Introduction of Metrics for Blockchain

Javier Díaz[1]([✉]) [iD], Mónica D. Tugnarelli[2] [iD], Mauro F. Fornaroli[2] [iD], Lucas Barboza[2],
Facundo Miño[2], and Juan I. Carubia Grieco[2]

[1] Facultad de Informática, Universidad Nacional de La Plata, 50 y 120, 1900 La Plata,
Buenos Aires, Argentina
jdiaz@unlp.edu.ar
[2] Facultad de Ciencias de La Administración, Universidad Nacional de Entre Ríos,
Av. Tavella 1400, 3200 Concordia, Entre Ríos, Argentina
monica.tugnarelli@uner.edu.ar

Abstract. This article presents an introduction to the definition of a set of metrics to address the performance, scalability and workload analysis of blockchain technologies, mainly with tools applied to the Argentine Federal Blockchain and Hyperledger Fabric. Issues to consider in the application of this technology for the protection of digital evidence are also raised. Although there are known methods to measure performance, there is still no common framework that facilitates the task of achieving a comparative measurement between the different blockchain solutions, which, considering the sustained use of this technology in a wide field of application, it is shown as a vacant area for which we consider that it is necessary to advance in order to evaluate the performance in different use cases and scenarios.

Keywords: Blockchain · Metrics · BFA · Ethereum · Hyperledger · Digital evidence

1 Introduction

This article presents the progress of the PID-UNER 7059 research called "Blockchain Technology for securing digital evidence in Forensic Readiness environments". Its main goal is to analyze the impact of the use of this technology when applied to the preservation, integrity and traceability of digital evidence. The latter is obtained *a priori* from the assets identified as essential in an organization, in a preventive environment such as Forensic Readiness, also called Forensic Preparation [1].

The PID has several stages, each of which is carried out in a computer laboratory using a blockchain with no associated cryptocurrency, to implement a prototype where tests can be carried out to analyze how this technology reacts to the Forensic Readiness requirements, both for the process of securing the evidence and for maintaining the chain of custody.

In previous works [2] the characteristics of different types of blockchain available in the market have been analyzed and, according to the objectives set out in this research, it

P. Pesado and G. Gil (Eds.): CACIC 2021, CCIS 1584, pp. 285–294, 2022.
https://doi.org/10.1007/978-3-031-05903-2_19

focuses the analysis on two representative solutions: a public, distributed and decentralized platform such as Ethereum [3] and a private solution with centralized administration, like Hyperledger Fabric [4]. This analysis considers aspects such as: privacy, security, transaction validation speed, use cases, open standards, among others [5].

This article extends what has been described in previous works [6] in terms of presenting preliminary metrics to measure the performance of BFA and Hyperledger. Some real-time-captured performance data is updated and the analysis of the consensus mechanisms used by both solutions is explained further. Finally, a first comparison of those features that may affect the environment of trust and privacy, required for the safeguarding of forensic evidence in the preventive environment Forensic Readiness, is proposed.

2   Metrics

As the PID stages progressed, it became necessary to have some indicators that help measure the performance and efficiency of each type of blockchain.

Although there are known aspects concerning performance measurement, there is no common framework that facilitates getting a comparative measurement in the different implementations of blockchain solutions, which, considering the uninterrupted use of this technology in different domains, is shown as a relevant area on which we consider it convenient to advance.

Hereupon, a metric will be the basic input to collect information about the behavior of the blockchain, which is why in this article some initials are introduced on the Argentine Federal Blockchain as an example of Ethereum and a first review of the topic on Hyperledger Fabric installed as a base of laboratory tests. For the first case, two tools that are available on the BFA site were used, *bfascan*[1] developed by the company Última Milla and a monitor implemented with Grafana[2] that has been put into operation by the Information Systems Agency GCABA. All of them which are presented next.

2.1   Metrics on BFA

The Argentine Federal Blockchain (BFA) [7] is an open and participatory multiservice platform based on Ethereum technology and designed to integrate services and applications on blockchain. It is currently integrated by 95 parties from public, private, academic and civil society sectors. All of them have equal participation in areas such as organizational engineering to infrastructure deployment where no sector has a majority and that prevents it from being manipulated.

It has a variety of use cases such as: verification of digital notarial documents; verification of university graduate records; certification suppliers offer receipts; exported fruit traceability; commercialization of agricultural commodities and validation of real estate sales tickets, to list a few.

Another resource provided by BFA is the Time Stamp 2.0 featured, that provides a secure official time to use in different processes. It makes it possible to prove the

---

[1] BFA SCAN bfascan.com.ar.
[2] Monitor https://bfa.ar/monitor.

existence of any digital document contents at a specific moment in time, which then allows to verify if future copies of it remained unchanged.

BFA uses Proof of Authority (PoA) with no associated cryptocurrency. The Proof of Authority model has certain number of nodes that are authorized to seal blocks, in this case 21 operating nodes, which do not compete with each other but take turns to add blocks to the chain. The BFA Consortium is the one who decides what new sealers can be added or removed, for what it requires the agreement and operation approval of (n/2) + 1 of its sealer nodes.

As there is no need to solve complex algorithms, the amount of processing is minimal, so these models are considered as light and more efficient in relation to energy consumption. The other interesting feature about BFA is that there is no circulation of cryptocurrencies with economic value, since a participation reward is not actually necessary nor pursued.

Therefore, each BFA entity that manages a node is responsible for its maintenance and monitoring and there is no central management system in the network. As supporting mechanism, BFA implements a monitoring scheme through the NOC (Network Operation Center), *"that will monitor the operation of the sealer and gateway acting nodes."* It does not have a centralized location. It is instead distributed geographically and among different regions of the country.

It worth noting that the Faculty of Administration Sciences at UNER is a member of the aforementioned organization and has also implemented a sealer node on its network, which is managed and supervised by two members of this research team.

Data was captured on these distributed nodes using available analysis tools.

First, the following table and graphs show data obtained on the Argentine Federal Blockchain in terms of the number of operating nodes, volume of transactions and accounts created in the BFA (Table 1):

Table 1. BFA data collection with BFA Scan as of 8/16/21. Source: self made

Number of nodes making up the network	-Total number of transactional nodes.	95 transactional nodes
	-Total number of sealing nodes	21 sealing nodes online and operational
Transactions of the day	Total number of transactions performed since 0:00hs	Variable information: average of 10,000 transactions on business days.
Total transactions	Total number of transactions performed.	Variable information: accumulated 5350650
Total Addresses	Total accounts created in the BFA.	Variable information: 805

Figure 1 shows the accumulated activity per month, which manifests that there are use peaks with a noticeable decrease on non-working days. It is also clear that its use volume is not elevated, maybe due to the fact that it is still a novel initiative that lacks notoriety and that has room for a more intense use and/or for more number of operations.

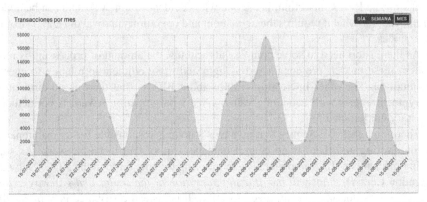

Fig. 1. Capture of monthly statistical information with BFAScan as of 8/16/21. Source: BFAScan website

Based on the information collected, it is proposed to define some initial metrics related to the workload, such as:

- **TPT_BFA**: Average time of a transaction = Transactions of the Day/elapsed time since 0:00 (at the moment of calculation).
- **CTPN_BFA**: Number of average transactions per node = Total Transactions/Total Sealing Nodes
- **CTND_BFA**: Average number of transactions per node of the day = Transactions of the Day/Total Sealing Nodes

In addition to the previous tool, the BFASCAN site offers an API called API REST BFA Scan that presents different methods to obtain information contained in the BFA.

a) **method getBlocks** url: http://201.190.184.52/bfascan/Blocks/getBlocks

- HTTP_GET. Returns information from the last 10 blocks.
- HTTP_POST. Returns information on the last 100 blocks, or one in particular (from the hash).
- HTTP_POST *for registered users*: Returns information on the last n blocks (maximum 1000), or one in particular (based on the hash). The query requires a token that is sent in an email after registration.

From the information returned by the getBlocks method, the timestamps of each record can be used to calculate the average creation time of a block. For example:

- **TPCB_BFA**: Average block creation time = time elapsed from the first to the last block returned by the query/the total number of blocks returned by the query. (The more blocks can be consulted, the more accurate the approximation can be).

Another field that the query returns on each record is transactions_associated. This value could be used to calculate the average number of transactions per block by adding the number of transactions in each block to the total number of blocks.

- **CPTB_BFA:** Average number of transactions per block = sum of the transactions associated with each of the blocks returned by the query/the total number of blocks returned by the query.

b) **getTxs method**: url http://201.190.184.52/bfascan/transactions/getTxs

- **HTTP_GET**: Returns information on the last 10 transactions created in the BFA.
- **HTTP_POST**: Returns information on the last 100 transactions, or one in particular (based on its hash)
- **HTTP_ POST** *for registered users*: Returns information on the latest transactions (maximum 1000), or one in particular (based on the hash). The query requires a token that is sent in an email after registration.

As with the getBlocks method, the timestamps of the records returned by the query could be used to calculate the average creation time of a transaction (or time between transactions). For example:

- **TPCT_BFA**: Average transaction creation time = time elapsed between the first and last returned transaction/number of returned transactions (same as in the previous case, the greater the number of records consulted, the more approximate the estimate could be)

These real-time queries must be repeated at regular intervals and then estimated from all the average times obtained.

Grafana's BFA monitor tool, and based on the information collected, allows obtaining information for other metrics:

- **CPBS_BFA**: Average number of sealed blocks per node (based on the information on the last sealed block in x period of time (allows queries for the last n minutes, hours, days) These values could be calculated at different times and then calculated an average over these for a better estimate.
- **TF_BFA**: Completion time: time required to achieve the immutability of transactions and blocks.

Figure 2 shows a real-time traffic capture with the detail of blocks per sealer node, the cost of each transaction and other monitoring data.

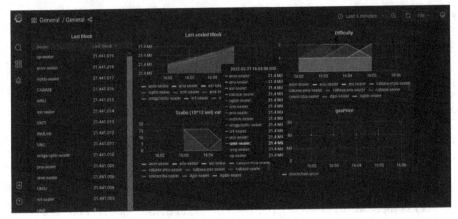

Fig. 2. Capture of monitoring information 2/27/22. Source: WebMonitor

2.2 Metrics on Hyperledger Fabric

Hyperledger Fabric is an open source distributed ledger technology (DLT) platform designed to be used in business contexts, which offers some key features, including: it is developed under the scope of the Linux Foundation; it is maintained by a set of members of 35 organizations; it has a modular and configurable architecture; it is the first distributed ledger platform to support smart contracts written in general purpose languages such as Java, Go and Node.js; the Fabric platform is authorized (permissioned), so the participants know each other, among others.

Hyperledger Fabric uses Kafka Orderer consensus mechanism that allows more than a thousand transactions per second. In this approach, authorized participants, with controlled access, validate transactions. When the majority validates a transaction, there is consensus and it is confirmed. This authorized voting strategy is advantageous as it provides low latency and good fault tolerance.

As an aspect to consider, while Kafka is fault tolerant, it is not against Byzantine faults, which could cause the system to not reach an agreement in the case of malicious or defective nodes. Billed as an improvement, later versions of Hyperledger have incorporated the leadership-based Raft consensus protocol, where "follower" nodes replicate the log entries created by the "leader" and can also elect a new leader node in the event that it stops sending messages after some configured time.

In this project we work with Hyperledger Fabric set up in a test laboratory. This implementation provides a file called *configtx.yaml* for configuration of certain network features. It contains two important parameters *BatchSize* and *BatchTimeout* that allow to configure the throughput and latency of transactions.

- **Batch size**: Defines how many transactions the computer node will collect before closing a block. No block will exceed the size of AbsoluteMaxBytes or have more than MaxMessageCount transactions inside. The ideal build block size is Preferred-MaxBytes. Transactions that are larger than this size will appear in their own block.
- **Batch timeout**: it is a reservation mechanism if the block is not filled in a specific time. This value provides an upper bound on the time it takes to close a block of transactions. Lowering this value will improve latency, but making it too small may reduce performance by not allowing the block to fill to its maximum capacity.

The lower the *Batch timeout* and *Batch size* values, the greater the total number of blocks generated per second. Instead, the higher their values, the lower the number of blocks generated per second.

Reducing the Batch timeout value will decrease latency, but at the expense of overall throughput. Conversely, increasing the *MaxMessageCount* will increase throughput but at the expense of transaction latency. The value of this latency, which is obtained by subtracting confirmation time and sending time, is a view of the entire network related to the amount of time it takes for a transaction to become effective and propagate throughout the network.

As an addition, different versions of Hyperledger Fabric (HLF), such as HLF v0.6 and HLF v1.0, should be compared in the same evaluation framework to demonstrate the performance advantages/disadvantages of new versions [8].

To continue with what is specified in the project, Hyperledger Caliper will be used in order to generate reports over a study time on different performance indicators, namely transactions per second, transaction latency, resource use and scalability, among others.

2.3 Considerations on Applicable Metrics for Chain of Custody Assurance

As mentioned in this article introduction, the main objective of the project is to analyze the use of blockchain for the preservation of the chain of custody based on the Forensic Readiness approach, where possible evidence must be protected before an incident occurs, to meet two objectives: maximizing the environment's ability to gather reliable digital evidence and minimizing the forensic cost during an incident response.

This chain of custody must be clearly documented and with a detailed traceability from its collection to its storage, which is why applying blockchain technology to fulfill this requirement is of special interest.

The following table provides a comparison of the main characteristics of the analyzed blockchains solutions, focusing on aspects that are relevant for the safeguarding of digital evidence (Table 2):

Table 2. Comparison of BFA versus Hyperledger Fabric.

Benefits	BFA	Hyperledger Fabric
Property of the environment	Ethereum-based, open source	Open source, Linux foundation
Authentication of participants	Consortium format made up of multiple independent parties where no party can control the system or alter the transfer record stored on the blockchain. Each party signs a commitment to participate (on paper) and submits supporting documentation validating its identity. Acts as a Permittee	It is a private permissioned network, with restricted access and the identity of the participants is known.
Availability	Decentralized.	Centralized.
Latency	Measured in minutes	Measured in seconds
Propagation of blocks	Dependent on network diameter and latency	High speed transaction processing
Transactions	All participants can access the transaction log	The registry is not public.
Integrity	SHA3 and others Timestamp 2.0	SHA 2 and others
Smart contracts	Use of Smart Contract for transactions	Use of Smart Contract for transactions
Consensus protocol robustness	Proof of Authority (PoA)	Proof of Authority (PoA)
Net diameter	Variable. Dynamic parts input and output	Can be fixed and restricted
Operating cost	It does not use cryptocurrency.	Does not use cryptocurrency
Hardware	Dedicated	Dedicated

3 Conclusions and Future Work

In this work, some preliminary metrics have been presented that will serve as the basis for measuring quality, performance and scalability of blockchain applications. The main benefit of having metrics is that they help identify and find possible problems in the operation of the blockchain, which in turn makes it possible to analyze decentralized and simultaneously acting nodes work. Metrics also help with identifying bottlenecks, determining the use of resources, optimizing consensus protocols and detecting network security attacks. Thus, metrics help create a more controlled and safer operating environment.

As future work, the proposed metrics will be used during a sufficient period of time to establish a test bench as a basis for analysis, both on the BFA blockchain and on the Hyperledger deployed in the laboratory, and make necessary adjustments. In addition, it is expected to progress by obtaining indicators related to security aspects in the blockchain, mainly by trying to measure the relationship between the implemented security scheme versus the functionalities that the blockchain must provide.

It is to be noted that the appropriate blockchain will be defined considering necessary security services for the protection of digital evidence and the chain of custody, both being aspects detailed as objectives of this work. In this regard, the technology that guarantees the integrity and non-repudiation of documents, audios and digital videos that can be evidence should be considered, that is, not only to observe the hash validation but also the protection of the transaction ID and the issuer to ensure the authentication of the participants involved with the evidence, of the timestamp so that digital copies of it are not generated and on the availability, for which tests related to attack targets must be carried out on a centralized distribution versus a decentralized one, added to inconveniences bottleneck that can function as a denial of service.

Future actions will also include the evaluation of some systems proposed by various authors [9–11], taking into account the complexity and challenges of safeguarding evidence that is volatile and the digital storage of forensic evidence as legal records. Achieving an environment of trust between the different actors of the judicial system is considered a strict and determining requirement that will affect the blockchain technology to be selected to meet the objectives of this project and that can be recommended for use.

References

1. Tan, J.: Forensic Readiness (2001). http://isis.poly.edu/kulesh/forensics/forensic_readiness.pdf
2. Díaz, F.J., Tugnarelli, M.D., Fornaroli, M.F., Barboza, L.: Blockchain para aseguramiento de evidencia digital en entornos Forensic Readiness. In: XXII Workshop de Investigadores en Ciencias de la Computación (WICC 2020) (2020). ISBN: 978-987-3714-82-5. http://sedici.unlp.edu.ar/handle/10915/103377
3. Ethereum. https://ethereum.org/en/. Accessed 3 Mar 2022
4. Hyperledger. https://www.hyperledger.org/use/fabric. Accessed 2 Mar 2022
5. Crosby, M., et al.: BlockChain technology: beyond bitcoin. Applied Innovation Review (AIR). Issue No. 2. Berkeley (2016). http://scet.berkeley.edu/wp-content/uploads/AIR-2016-Final-version-Int.pdf
6. Díaz, F.J., Tugnarelli, M.D., Fornaroli, M.F., Barboza, L., Miño, F.: Métricas para blockchain. In: XXVII Congreso Argentino de Ciencias de la Computación (CACIC 2021) (2021). ISBN: 978-987-633-574-4. http://sedici.unlp.edu.ar/handle/10915/129809
7. Blockchain Federal Argentina. https://bfa.ar/. Accessed 3 Mar 2022
8. Fan, C., Ghaemi, S., Khazaei, H., Musilek, P.: Performance evaluation of blockchain systems: a systematic survey. IEEE Access 8, 126927–126950 (2020). https://doi.org/10.1109/ACCESS.2020.3006078.(2020)
9. Verma, A., Bhattacharya, P., Saraswat, D., Tanwar, S.: NyaYa: blockchain-based electronic law record management scheme for judicial investigations. J. Inf. Secur. Appl. 63, 103025 (2021). ISSN 2214-2126. https://doi.org/10.1016/j.jisa.2021.103025
10. Guo, J., et al.: Antitampering scheme of evidence transfer information in judicial system based on blockchain. Hindawi. Secur. Commun. Netw. 2022 (2022). https://doi.org/10.1155/2022/5804109
11. Lone, A.H., Mir, R.N.: Forensic-chain: blockchain based digital forensics chain of custody with PoC in hyperledger composer. Digit. Invest. (2019). https://doi.org/10.1016/j.diin.2019.01.002

12. Nakamoto, S.: Bitcoin: a peer-to-peer electronic cash system (2009). https://bitcoin.org/bitcoin.pdf
13. Yang, D., Long, C., Xu, H., Peng, S.: A review on scalability of blockchain. In: Proceedings of the 2020 The 2nd International Conference on Blockchain Technology, pp. 1–6. Association for Computing Machinery, New York (2020). https://doi.org/10.1145/3390566.3391665
14. Tuan, T., et al.: BLOCKBENCH: a framework for analyzing private blockchains. https://www.comp.nus.edu.sg/~ooibc/blockbench.pdf (2017)
15. Ballandies, M.C., Dapp, M.M., Pournaras, E.: Decrypting distributed ledger design—taxonomy, classification and blockchain community evaluation. Cluster Comput. (2021). https://doi.org/10.1007/s10586-021-03256-w

Digital Governance and Smart Cities

Data Quality Applied to Open Databases: "COVID-19 Cases" and "COVID-19 Vaccines"

Ariel Pasini[1,2,3]($\boxtimes$), Juan Ignacio Torres[1,2,3], Silvia Esponda[1,2,3], and Patricia Pesado[1,2,3]

[1] Computer Science Research Institute LIDI (III-LIDI), La Plata, Argentina
{apasini,jitorres,sesponda,ppesado}@lidi.info.unlp.edu.ar
[2] Facultad de Informática, Universidad Nacional de La Plata, 50 y 120, La Plata, Buenos Aires, Argentina
[3] Centro Asociado Comisión de Investigaciones Científicas de La Pcia. de Bs. As. (CIC), Tolosa, Argentina

Abstract. To increase transparency, encourage citizen participation in decision making, and to respond efficiently, governments make a very important set of information available to their community. This information became even more relevant during COVID-19. However, if these data do not have a high level of quality, information loses reliability. An evaluation to measure data quality of two public files was conducted: "COVID-19. Cases registered in the Argentine Republic." and "COVID-19 vaccines. Doses administered in the Argentine Republic." using the model provided by the ISO/IEC 25012 standard and the evaluation process defined by ISO/IEC 25040.

Keywords: Data quality · ISO/IEC 25000 · Open government · Open data

1 Introduction

Since the first case detected in Wuhan (Hubei, China), COVID-19 has spread rapidly over approximately 200 countries/regions threatening lives and significantly disrupting the global economy and society.

Publishing open data is crucial during pandemics, since, through their analysis and processing, citizens can participate in the decision-making process to respond intelligently [1]. It is essential that data published in open formats (either through datasets or daily reports) is of high quality. Low data quality negatively affects the value creation process and transparency and trust in government [2].

The general model defined in the ISO/IEC 25012 standard – "Data Quality Model" [3] allows data quality measurement of data stored in a computer system. It is used in conjunction with other standards to plan and conduct data quality evaluations. This model classifies quality attributes into fifteen characteristics analysed from two points of view which are not mutually exclusive: inherent and system-dependent. These characteristics obtain different importance and priority according to each evaluator based on their own specific needs.

P. Pesado and G. Gil (Eds.): CACIC 2021, CCIS 1584, pp. 297–311, 2022.
https://doi.org/10.1007/978-3-031-05903-2_20

On the other hand, the ISO/IEC 25040 standard – "Evaluation Process" [4] provides a general reference model for evaluation, which considers the inputs to the evaluation process, restrictions and resources that are necessary to obtain the required outputs. The process is composed of five activities.

This article proposes to measure the quality of the data obtained from two public files [5], maintained by the Ministry of Health of Argentina, selecting certain quality characteristics defined by the ISO/IEC 25012 standard and following the activities provided in the evaluation model established by the ISO/IEC 25040. The contributions of this work can be seen as an extension of [6]. Besides considering the public file *"COVID-19. Cases registered in the Argentine Republic"*, this work incorporates the data from *"COVID-19 vaccines. Doses administered in the Argentine Republic"*. Even more, a comparison between both resources is also addressed.

The second section introduces the standards and metrics to be used for data quality assessment. In section three, the concept of Open Government is introduced, advantages and disadvantages of publishing open data are mentioned and the importance of citizen participation in the value creation process using these data is highlighted. Then, in the fourth chapter the resources to be evaluated are introduced in detail and the quality evaluations of the databases are carried out, as well as a comparison between both resources. Finally, the last chapter presents the conclusions of the article.

2   Met Database Quality

The ISO/IEC 25000 family, known as SQuaRE (System and Software Requirements and Evaluation) intends to assess the quality of a software product in a common work framework. In the context of the article, two regulations of the family will be used: ISO/IEC 25012 and ISO/IEC 25040.

2.1   ISO/IEC 25012 Database Quality Model

Undoubtedly, the amount of data run by computer systems is increasing worldwide. It is necessary, in every information technology project, to maximize the data quality that is interchanged, processed, and used among systems. A low amount of data can generate unsatisfactory information and useless results.

The standard defines a data quality model in a structured model within a computer system. This model is composed of fifteen characteristics divided into two points of view: system-inherent and system-dependent. It should be noted that some characteristics are relevant to both points of view.

Inherent Data Quality
This point of view refers to the level in which the characteristics have the potential to satisfy the established and implicit needs when data is used under specific conditions. The relevant characteristics from the inherent point of view are: *Accuracy*: it makes reference to the fact that data has attributes represented in a correct way (it can be syntactic or semantic). *Completeness*: data has values for all the expected attributes *Consistency*: data is not contradictory and is consistent. *Credibility*: data has accurate and credible

attributes (it includes authenticity concept). *Currentness*: data has attributes that are updated properly.

System-Dependent Data Quality

It makes reference to the level of data quality that is achieved and preserved in a computer system when data is used under specific conditions. There exists a dependence between the data quality and the technological domain in which they are used. The relevant characteristics from the system-dependent point of view are: *Availability*: it defines the level in which attributes can be recovered by the users or applications that can get access to them. *Portability*: level in which data can be installed, replaced or copied from a system to another keeping its quality. *Recoverability*: data must keep and preserve a special level of operations and quality in the presence of failures.

Inherent and System-Dependent Data Quality

The relevant characteristics from both points of view (inherent and System-dependent) are: *Accessibility*: level in which data can be accessed. *Compliance*: data fulfills regulations, conventions or standards related to data quality in a specific context. *Confidentiality*: data has attributes that must be interpreted and accessed by authorized users. It is part of information security along with integrity and availability. *Efficiency*: attributes can be processed and, therefore, provide expected levels of efficiency using types and quantities of corresponding resources. *Precision*: attributes that belong to data present accurate values or values that can be differentiated. *Traceability*: it is analysed whether a register for the audit of data access is provided or not and any modification regarding them. *Understandability*: level in which users can read and interpret data. They must be expressed clearly.

2.2   Metric Definitions According to ISO/IEC 25012

The ISO/IEC 25012 gives metrics as example for each characteristic. Therefore, for each characteristic, the measurement functions in the regulation will be used. It is presented one as an example:

- Type (inherent or system-dependent)
- Name of the quality measure
- Measurement function
- Variable

In Table 1 the metric for the characteristic Completeness is presented as an example.

Table 1. Metric definition for completeness

Type	Inherent
Name of the quality measure	Completeness of data in a file
Measurement function	Value = A/B
Variables	A = number of data required for the particular context in the file B = number of data in the particular context of expected use

2.3 Assessment Model

The ISO/IEC 25040 regulation defines a reference model taking into account necessary inputs, restrictions and resources so as to obtain the required outputs. In order to carry out the assessment of a software product, a process of five activities is defined:

1. **To establish the assessment requirements**: to establish the purpose of the assessment and the software quality requirements to be assessed by identifying the interested parts in the product, the risks (if there are any) and the quality model to be used.
2. **To specify the assessment**: within this activity the assessment modules are selected and the decision criteria are defined both for the metrics and for the assessment.
3. **To design the assessment**: the plan with the assessment activities that must be carried out is defined, taking into account the resources availability.
4. **To carry out the assessment**: in this fourth activity, the measurements are carried out and the decision criteria is used both for the metrics and for the assessment.
5. **To conclude the assessment:** the quality assessment is closed. Based on what it was obtained, a report with the final results and the conclusions of the assessment are carried out.

3 Government and Open Data

The term 'open government' was being built through time. Nevertheless, it was popularized in 2009 by the "Memorandum on Transparency and Open Government" [7] during the Barack Obama presidency, where three principles were proposed:

1. A government must be transparent. Transparency in a government can be achieved by giving citizens information about what they are carrying out. This promotes accounting reports.
2. A government must be participatory. The public participation in the politics formation as well as the contribution of knowledge, ideas and experience improve the government effectiveness and the quality of their decisions.
3. A government must be collaborative. Collaboration involves citizens actively in their government work. This principle requires cooperation among individuals, companies, organizations and agents in all the government levels.

Open government can be understood as a technological platform, since it uses information and communication technology with all the available means to achieve its goal. This includes both websites and social networks [8].

3.1 Participation of Citizens in the Analysis of Open Data

Participation of citizenship is crucial in the concept of open government. Governments around the world look for the increase of citizenship participation due to the importance in the decision making process, both in political and administrative processes. Therefore, it is essential to bring public data in open format closer to society. The TIC revolution (communication and information technology) changed radically the interaction between governments and citizens.

Social Web or 2.0 Web, which encourages the participation of individuals, can easily reach the fulfillment of the three goals mentioned in the previous part. This paradigm is about the content that is created, shared and processed by users through media for social communication and the creation of a social network in which they can be connected one another [9].

Governments can expand the scope of open data that they publish through the use of such electronic social networks. In order to promote this, it is necessary to create communities around data, that they can be consulted and analysed, as well as sharing new data. Furthermore, a key point to bear in mind is their redistribution: if the redistribution is fast, the value obtained by data will be higher.

3.2 Open Data: Advantages and Disadvantages

The publication of open data in their different ways, as it was mentioned previously, allows citizens to use and create information through a collaborative network. With such network, a collaborative intelligence arises in which the public helps to better the decision-making process by participating actively.

Also, the promotion of the open data publication, increases people's satisfaction, since it encourages equality to data access and this builds up confidence and transparency [10].

However, there are certain factors that are detrimental to the data use and participation of the citizenship since they do not encourage the creation of the added value. Another concern is the lack of a systematic analysis about what should be published and what is expected by the users about open data. In many cases, the lack of analysis involves the publication of incomplete- or even excessive- information, so concentrating on what is relevant is impossible. Therefore, it is important to guarantee the data quality in order to ensure a proper analysis.

4 COVID-19 Open Databases

Publishing open epidemiological data together with its analysis is a fundamental part to fight COVID-19. That is why, to be more transparent during this pandemic, the National Ministry of Health published different datasets related to the fight against coronavirus. This article was focused on the following files: *"COVID-19. Cases registered in the*

Argentine Republic" and *"COVID-19 vaccines. Doses administered in the Argentine Republic"* maintained by the National Ministry of Health, which are daily updated and began to be published on May 15, 2020, and March 23, 2021, respectively.

4.1 COVID-19 Cases in the Argentine Republic Dataset

"COVID-19. Cases registered in the Argentine Republic" is a dataset which has twenty-five fields, which are listed along with their types: case number (integer), sex (string), age (integer), age in months/years (string), country of residence (string), province of residence (string), department of residence (string), province in which the data was loaded (string), date of onset (date according ISO-8601 - "Data elements and interchange formats—Information interchange—Representation of dates and times"), case opening date (ISO-8601 date), epidemiological week of opening date (integer), hospitalization date (ISO-8601 date), intensive care required (string), date of admission to intensive care if applicable (ISO-8601 date), death (no/yes) (string), date of death if applicable (ISO-8601 time), mechanical ventilation required (string), province in which the data was loaded id (integer), funding source (string), manual registry classification (string), case classification (string), province of residence id (integer), diagnosis date (ISO-8601 time), department of residence id (integer) and last update (ISO-8601 date).

4.2 COVID-19 Vaccines in the Argentine Republic Dataset

"COVID-19 vaccines. Doses administered in the Argentine Republic" has the following sixteen fields: sex (string), age group (string), jurisdiction of residence (string), jurisdiction of residence id (integer), department of residence (string), department of residence id (integer), jurisdiction where the dose was given (string), jurisdiction where the dose was given (id) (integer), department where the dose was given (string), department where the dose was given (code) (integer), date of administration (ISO-8601 date), vaccine (string), condition of administration (string), dose order (integer), dose name(string) – later removed and batch of vaccine (string).

4.3 COVID-19 Databases Evaluation

The evaluation of both datasets will be carried out considering the activities proposed by the ISO/IEC 25040 standard.

4.3.1 Purpose of Evaluation

The purpose of the evaluation is: *to analyse whether the data considered mandatory is not empty, whether the classification of cases and the vaccine administrations are up-to-date, whether the fields comply with the expected formats, whether they are expressed correctly and if the data is consistent per se.*

To achieve this, the following characteristics will be evaluated: **Accuracy, Completeness, Consistency, Currentness** and **Understandability**.

4.3.2 Evaluation Specification

In this activity, in addition to the fact of defining the evaluation modules for both databases (presented in 2.2), the decision criteria are defined heuristically for both the characteristics and the final evaluation.

Table 2 shows the range of values to determine the level of acceptance for these characteristics: *accuracy, consistency, completeness, and understandability).*

In the case of currentness, Table 3 presents the decision criteria for it in the database of dose administration, while in the registered cases dataset, in order to determine the level obtained, three fields must be verified: confirmed, deaths and negatives, Table 4 shows the values to determine the level of each one of them, and finally Table 5 describes the combination of all three fields to establish the level reached by the characteristic.

Table 2. Decision criteria for accuracy, consistency, completeness and understandability.

Characteristics	Level	Range
Currentness Consistency Completeness Understandability	Unacceptable	If Value $>= 0$ and Value $< 0,3$
	Minimally acceptable	If Value $>= 0,3$ and Value $< 0,7$
	Target range	If Value $>= 0,7$ and Value $< 0,9$
	Exceeds requirements	If Value $>= 0,9$

Table 3. Decision criteria for currentness in the vaccine administration dataset.

Characteristic	Level	Range
Currentness	Unacceptable	If Value $>= 0$ and Value $< 0,3$
	Minimally acceptable	If Value $>= 0,3$ and Value $< 0,7$
	Target range	If Value $>= 0,7$ and Value $< 0,9$
	Exceeds requirements	If Value $>= 0,9$

Table 4. Decision criteria for the fields in the registered cases dataset.

Campos	Level	Range
Confirmed (C) Deaths (D) Negative (N)	Unacceptable	If Value $>= 0$ and Value $< 0,3$
	Minimally acceptable	If Value $>= 0,3$ and Value $< 0,7$
	Target range	If Value $>= 0,7$ and Value $< 0,9$
	Exceeds requirements	If Value $>= 0,9$

Table 5. Decision Criteria for Currentness in the registered cases dataset considering the three fields.

Characteristic	Level	Range
Currentness	Unacceptable	C, D and N: Unacceptable
	Minimally acceptable	C, D and N: Minimally acceptable
	Target range	C, D and N: Target range
	Exceeds requirements	C, D and N: Exceeds requirements

Table 6 shows the decision criteria for the final evaluation, considering the decision criteria for each characteristic.

Table 6. Decision criteria for the final evaluation.

Final level	Value of characteristics
Unacceptable	A, C, R and U: Unacceptable O: Minimally acceptable
Minimally acceptable	A, O and R: Minimally acceptable C and U: Target range
Target range	A, O, R and U: Target range C: Exceeds requirements
Exceeds requirements	A, C, O, R and U: Exceeds requirements

Accuracy (A) Completeness (C) Consistency (O) Currentness (R) Understandability (U).

4.3.3 Assessment Design

A data quality assessment of both datasets is carried out, from their publication date (May 15, 2020, for the registered cases and March 23, 2021 for the vaccine administration) until January 31, 2022. The Accuracy, Completeness, Consistency, Currentness and Understandability (defined in the ISO/IEC 25012 standard) are evaluated, based on the guide proposed by ISO/IEC 25040 and then the average is calculated to determine the final value and level. The specifications to carry out the evaluation of the registered cases are shown in Table 7. In the case of vaccine doses, the specifications are indicated in Table 8.

Table 7. Specifications of each characteristic to be evaluated for the registered cases

Characteristic	Specifications
Accuracy (Syntactic + Semantic)	Each field must match the data type defined by the database administrator in the context according to the description Case number → Integer Sex → String (M o F) Age → Integer between 0 and 116 Age in months/years → String ("years" or "months") Country of residence → String Province of residence → String Department of Residence → String Province in which the data was loaded → String Date of onset → ISO-8601 Date (date) Case opening date → ISO-8601 Date (date) Epidemiological week of opening date → Integer Hospitalization date → ISO-8601 Date (date) – if applicable Intensive care required → String (YES/NO) Date of admission to intensive care → ISO-8601 Date (date) – if applicable Death → String (YES/NO) Date of death → ISO-8601 Time (time) – if applicable Mechanical ventilation required → String (YES/NO) Province in which the data was loaded (id) → Integer Funding source → String (Public/Private) Manual registry classification → String Case classification → String (Negative, confirmed or possible) Province of residence (id) → Integer Diagnosis date → ISO-8601 Time (time) Department of residence (id) → Integer Last update → ISO-8601 Date (date)
Completeness	In each record it is evaluated that mandatory fields are not empty
Consistency	Data must be consistent. An example of inconsistency to consider is that a person who has not died has a date of death
Currentness	Currentness is measured based on the difference between the publication date of the dataset and the date of diagnosis (discarded/confirmed) or date of death. To carry out the assessment, what is considered current is the diagnosis date or date of death which have a maximum delay of 2 days with respect to the publication date of the file
Understandability	Fields must be expressed in an appropriate unit or form and must be readable by the user

Table 8. Specifications of each characteristic to be evaluated for the vaccines dataset

Characteristic	Specifications
Accuracy (Syntactic + Semantic)	Each field must match the data type defined by the database administrator in the context according to the description Sex → String (M o F) Age group → String (Grouped by ten years) Jurisdiction of residence → String Jurisdiction of residence (id) → Integer Department of residence → String Department of residence (id) → Integer Jurisdiction where the dose was given → String Jurisdiction where the dose was given (id) → Integer Department of administration → String Department of administration (id) → Integer Date of administration → ISO-8601 Date (date) Vaccine → String Condition of administration → String Dose order → Integer Dose name → String Batch of vaccine → String
Completeness	In each record it is evaluated that mandatory fields are not empty
Consistency	Data must be consistent. An example of inconsistency to consider is that a vaccine was administered before the date of the first registered vaccine
Currentness	Currentness is measured based on the difference between the publication date and the date of dose administration
Understandability	Fields must be expressed in an appropriate unit or form and must be readable by the user

4.3.4 Execution of the Evaluation

Measurements were made for both datasets, therefore the values will be obtained in order to apply later the decision criteria and determine the level of each characteristic.

4.3.4.1 Execution of the Quality Assessment for the Cases Dataset

After measurements, the values for each characteristic were obtained and the results of applying the decision criteria are reflected in Table 9.

Table 9. Values and results for each characteristic

Characteristic	Value	Result
Accuracy	1	Exceeds requirements
Completeness	0.83	Target range
Consistency	1	Exceeds requirements
Currentness	Confirmed = 0.76	Target range
	Negatives = 0.68	Minimally acceptable
	Deaths = 0.37	Minimally acceptable
Understandability	0.92	Exceeds requirements

The lowest result is the one obtained by the *deaths* field of the *currentness* characteristic, the average delay (in days) was 3.82 for *confirmed*, 16.13 days for *negatives* and 16.56 days for *deaths*.

4.3.4.2 Execution of the Quality Assessment for the Dose Administration Dataset

Same measurements were made for the vaccines dataset, and the values for each characteristic were obtained. Table 10 shows the results of applying the decision criteria for each characteristic.

Table 10. Values and results for each characteristic

Characteristic	Value	Result
Accuracy	1	Exceeds requirements
Completeness	0.98	Exceeds requirements
Consistency	1	Exceeds requirements
Currentness	0.95	Exceeds requirements
Understandability	0.92	Exceeds requirements

In this case, all the values are above 0.90, therefore, all characteristics exceed requirements.

4.3.4.3 Comparative Analysis Between Both Datasets

It is highly important to make a comparison between both datasets. Figure 1 shows the fatality rate (% of deaths divided by the number of cases) during the analysed period.

It also shows jumps, which reflect loading problems. For example, starting in early May 2020, delayed negatives began to be uploaded, and this prevented the case fatality rate from being accurate for a few months, until it stabilised. On October 3, 2020, there was another jump in the case fatality rate, due to delayed loading of deaths. Finally, the last loading problem was at the end of the year 2021, in which there was also a loading delay in the negatives, which was solved at the beginning of 2022.

The first vaccination record (700 doses) was on December 29, 2020, we can see the effectiveness of vaccination in Fig. 1. It shows how the fatality rate begins to decrease from the starting date of vaccination as the percentage of people vaccinated with at least one dose increases (Fig. 2).

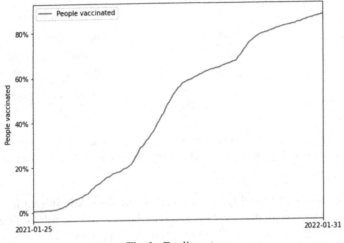

Fig. 1. Fatality rate

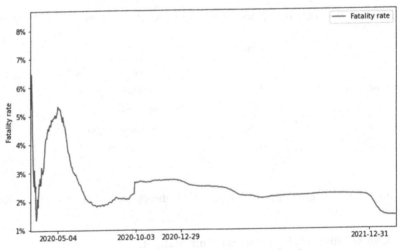

Fig. 2. People who took at least one vaccine dose.

4.3.5   Conclusion of the Evaluation

Tables 11 and 12 show the results obtained for each characteristic in the datasets of registered cases and dose administration, respectively, applying the decision criteria defined for the final evaluation (Table 5).

In the case of registered cases, although three characteristics exceed requirements, the currentness of deaths and negatives determined that the *currentness* characteristic is minimally acceptable, therefore, the highest level that can be granted to the final evaluation is: **Minimally acceptable**, being this a satisfactory result as defined in the ISO/IEC 25040 standard.

For the dose administration dataset, all characteristics exceed requirements, consequently, the result of the evaluation **Exceeds requirements** is also satisfactory according to ISO/IEC 25040.

Table 11. Final results (registered cases)

Characteristic	Level
Accuracy	Exceeds requirements
Completeness	Target range
Consistency	Exceeds requirements
Currentness	Minimally acceptable
Understandability	Exceeds requirements

In terms of understandability, although it exceeds requirements, there was no difference in the use of the ISO-8601 date (date) and ISO-8601 time (time), the fields with these types showed data with the same format.

Table 12. Final results (dose administration)

Characteristic	Level
Accuracy	Exceeds requirements
Completeness	Exceeds requirements
Consistency	Exceeds requirements
Currentness	Exceeds requirements
Understandability	Exceeds requirements

5   Conclusions

Two regulations of the ISO/IEC 25000 (SQuaRE) family were presented. The fifteen characteristics present in ISO/IEC 25012 were shown, five of these were selected for a specific context of use and the measures based on the measurement functions defined in the regulation were defined.

The concept of open government was defined, emphasising benefits and problems of the publication of open data, as well as highlighting the importance of the participation of the residents in it, in order to generate values in the society and to make the decision making process effective.

Data quality assessments were carried out to the following datasets: *"COVID-19. Cases registered in the Argentine Republic"* and *"COVID-19 vaccines. Doses administered in the Argentine Republic"* following the assessment model proposed by the ISO/IEC 25040 regulation. The final result for the cases registered was "minimally acceptable" due to the fact that the obtained value for the Currentness of deaths and discarded was within this range. In the case of vaccine doses administered, the final result was "exceeds requirements", since its five characteristics were at this level. It should be pointed out that both results were satisfactory.

The use of the standards mentioned allows us to perform a quantitative analysis of the data quality level, in this case in particular, of the datasets *"COVID-19. Cases registered in the Argentine Republic"* and *"COVID-19 vaccines. Doses administered in the Argentine Republic"*.

Both resources were compared by means of a graphic of death rate, showing some load problems and proving the effectiveness of the vaccination, while the number of vaccinated people, with at least one dose, increases.

Acknowledgements. This publication was made in the context of the CAP4CITY Project – "Strengthening Governance Capacity for Smart Sustainable Cities" (www.cap4city.eu) co-funded by the Erasmus+ Programme of the European Union. Grant no: 598273-EPP-1- 2018-1-AT-EPPKA2-CBHE-JP. Project number: 598273

References

1. Morgan, O.: How decision makers can use quantitative approaches to guide outbreak responses. Philos. Trans. R. Soc. B Biol. Sci. **374**(1776) (2019). https://doi.org/10.1098/rstb.2018.0365
2. Chen, H., Hailey, D., Wang, N., Yu, P.: A review of data quality assessment methods for public health information systems. Int. J. Environ. Res. Public Health **11**(5), 5170–5207 (2014). https://doi.org/10.3390/ijerph110505170
3. ISO/IEC 25012:2008: Software engineering—Software product Quality Requirements and Evaluation (SQuaRE)—Data quality model
4. ISO/IEC 25040:2011: Systems and software engineering—Systems and software Quality Requirements and Evaluation (SQuaRE)—Evaluation process
5. COVID-19: Casos registrados en la República Argentina, Vacunas contra COVID-19. Dosis Aplicadas en la República Argentina. https://datos.gob.ar/. http://datos.salud.gob.ar/
6. Pasini, A., Torres, J.I., Esponda, S., Pesado, P.: Calidad de datos aplicada a la base de datos abierta de casos registrados de COVID-19. In: Congreso Argentino de Ciencias de la Computación – CACIC 2021, pp. 735–745 (2021). ISBN: 978-987-633-574-4
7. Open Government: Memorandum on Transparency and Open Government. Fed. Regist., 21–22 (2009). https://www.archives.gov/files/cui/documents/2009-WH-memo-on-transparency-and-open-government.pdf

8. Sandoval-Almazán, R.: Gobierno abierto y transparencia: Construyendo un marco conceptual. Convergencia **22**(68), 203–227 (2015). https://doi.org/10.29101/crcs.v0i68.2958
9. Chun, S.A., Shulman, S., Sandoval, R., Hovy, E.: Government 2.0: making connections between citizens, data and government. Inf. Polity **15**(1–2), 1–9 (2010). https://doi.org/10. 3233/IP-2010-0205
10. Janssen, M., Charalabidis, Y., Zuiderwijk, A.: Benefits, adoption barriers and myths of open data and open government. Inf. Syst. Manag. **29**(4), 258–268 (2012)

Lexical Analysis Using Regular Expressions for Information Retrieval from a Legal Corpus

Osvaldo Mario Spositto(iD), Julio César Bossero(iD), Edgardo Javier Moreno(iD),
Viviana Alejandra Ledesma(✉)(iD), and Lorena Romina Matteo(iD)

Department of Engineering and Technological Research, National University of La Matanza,
Florencio Varela 1903, San Justo, La Matanza, Buenos Aires, Argentina
{spositto,jbossero,ej_moreno,vledesma,lmatteo}@unlam.edu.ar

Abstract. This article presents part of the work carried out in the framework of a research that aims to optimize an Information Retrieval System, by means of its specialization for the retrieval of legal documents. One of the fundamental sub-processes in this type of system is lexical analysis, in which indexing techniques are applied. These techniques involve extracting a series of concepts representative of the topics covered in a document, and then using them as access points for retrieval. This article describes a proposal for the extraction of information and identification of dates and references to named entities, such as File No., Resolution No., Article No. of Law XXX, which refer to the legal norm in force and are widely used in different judicial documents. For the recognition of such named entities, the process employed the definition of patterns using Regular Expressions, a way of representing a language in a synthetic form, applying a set of rules. From this, the terms obtained are stored in a matrix of terms/documents. This paper also describes the algorithms used during the validation of the proposed solution and presents the experimental results that show that by applying this method a significant reduction in the size of the inputs to the matrix can be achieved.

Keywords: Information retrieval systems · Regular expressions · Recognition of named entities · Lexical analysis

1 Introduction

This article is a continuation of the work presented at the 26th Argentine Congress of Computer Science, CACIC 2021, held in the city of Salta from 4 to 8 October 2021, organized by the Network of National Universities with Computer Science Degrees (RedUNCI) and the National University of Salta, under the title *"Propuesta para la construcción de un corpus jurídico utilizando Expresiones Regulares"* (Proposal for the construction of a legal corpus using Regulatory ExpressionsResults obtained) [1]. In this work, a theoretical proposal was presented to incorporate, in a legal corpus, references to dates and other common terms regularly used in the legal norm, by means of the Named Entity Recognition (NER) that make up the different judicial documents, using Regular Expressions (RE). RE are character strings that are used to describe or find patterns within other texts, using delimiters and syntax rules.

P. Pesado and G. Gil (Eds.): CACIC 2021, CCIS 1584, pp. 312–324, 2022.
https://doi.org/10.1007/978-3-031-05903-2_21

As expressed in the predecessor paper, jurisprudence has an important role as a source of law, because its conclusions support the application of the law in a specific case. The Argentinean judiciary produces every year a large number of rulings, files, among other things. These decisions are stored in documents, making this source of law ever larger, which drives professionals to spend more time to find relevant documents. Therefore, sophisticated computing techniques are needed to minimize search time and improve the relevance of retrieved documents.

The authors of this report belong to a group, which in the year 2021 presented a research project called "*Implementation of a Web System for Information Retrieval Oriented to Legal Documentation with the Parallelized Latent Semantic Indexing Process*", through the Incentive Program for Research Teachers of the Secretariat of University Policies (PROINCE).

Within the stages to carry out the aforementioned project, the theory and practice of documentary analysis, conceptual indexing, the development of a matrix of terms and the term/document matrix, which presents rows that correspond to terms and columns to documents, in this case a vector of documents is represented as a bag of words. In other words, to represent the textual content of the documents, this proposal uses a data structure consisting of a matrix with two dimensions, in which the indexing terms that have been extracted after processing the documents are stored in the system. These matrixes are widely used in the area of information retrieval, where the bag-of-words hypothesis captures to some extent the subject matter of the document [2].

As mentioned above, this article presents, on the one hand, the proposal to incorporate the references of both dates and the legal norm, through the NER or extraction of entities, such as Files, Decrees, Agreements, Articles, Laws, and others, that make up the different judicial documents, by means of patterns defined by RE. As an extension, an experimental test is also described in order to validate the proposed solution with the concrete results obtained.

2 Research Background

Regarding the work on corpus construction using RE to solve named entities, the work developed by Haag, in his thesis: "*Recognition of named entities in legal domain text*" [3], focuses on the detection, classification and annotation of named entities (e.g., Laws, Resolutions or Decrees) for the InfoLEG[1] corpus, a database that contains the documents of all the laws of the Argentine Republic. It should be noted that the search pattern presented in this article is based on the format presented by Haag, although it should be noted that his work does not include dates.

For his master's thesis, Duque Bedoya [4], presents a methodology to build a corpus for linguistic analysis, implementing a tagging system that includes the names of people, places, and dates by means of a computational tool. The information extraction processes include the automatic identification of such terms through the application of algorithms and heuristics used in digital libraries. The identification of events is carried out using the combination of the tags previously extracted from the corpus. The tool used for tagging

---

[1] http://www.infoleg.gob.ar/.

was Unstructured Information Management Architecture (UIMA) a free application to implement linguistic resources using JAVA and C++ programming languages, being compatible with the Eclipse programming environment.

Regarding the construction of various corpora, in the doctoral thesis of Rodriguez Inés [5], "*El uso de corpus electrónicos para la investigación de terminología jurídica*", an extensive list of the corpora available in Argentina and a detailed description on of more than 10 international multilingual corpora can be found. In addition, Cardellino's paper [6], "*A Low-cost, High-coverage Legal Named Entity*", can be mentioned. In this paper, an attempt is made to improve information extraction in legal texts by creating a legal named entity recognizer, classifier, and linker. The resulting tools and resources are open source and are aimed at developing a named entity recognizer, classifier and linker that exploits Wikipedia.

Another work worth mentioning is found in chapter two "*Regular Expressions, Text Normalization, Edit Distance*" of Jurafsky and Martin's book "*Speech and Language Processing*" [7], where a very clear explanation of the use of RE is given. In addition, a tool for performing language processing using RE is introduced in a theoretical way, and defines how to perform basic text normalization tasks, including word segmentation and normalization, sentence segmentation and lemmatization.

Finally, Robaldo et al., in their paper "*Compiling Regular Expressions to Extract Legal Modifications, present a prototype to automatically identify and classify types of modifications in Italian legal text*" [8]. This prototype uses XML language to define a new set of rules to identify the type of modifications in a text. The rule-based semantic interpreter implements a RE-based pattern matching strategy.

The proposal being presented in this paper is inspired by several aspects of the literature consulted and implements them in a process where a term matrix including the NER and date formats is created.

3   About Information Retrieval Systems

An Information Retrieval System (IRS) [2, 9, 10] is a tool that interacts between a corpus and its users. Its effectiveness depends on the adequate control of the language of representation of the information elements and the searches of its users. To meet its objectives, according to Tolosa et al. [2], an IRS must perform the following basic tasks:

- Logical representation of documents and, optionally, storage of the original.
- Representation of the user's need for information in the form of a query.
- Evaluation of the documents with respect to a query to establish the relevance of each one.
- Ranking of the documents considered relevant to form the "solution set" or response.
- Presentation of the response to the user.
- Feedback of the queries to increase the quality of the answer.

Robredo in [11], states that in any area of knowledge, meaningful terms can be used as descriptors to represent the content of written documents, in the processes of indexing and organizing information, as well as to formulate questions in the information retrieval process. Tolosa et al. in [2] state that the process can be divided into the following stages:

- Lexicographic analysis, words are extracted and normalized.
- Reduction (tokenization) of empty or high-frequency words.
- Lemmatization, words morphologically similar to a base or root form are reduced in order to increase the efficiency of a CRS.
- Selection of the terms to be indexed. Those simple or compound words that best represent the content of the documents are extracted.
- Assignment of weights or weighting of the terms that make up the indexes of each document.

With the advance of technology, the different communities generate an increasing volume of publications in different domains that are rapidly disseminated through different repositories. In this context, in order for users to find relevant publications for various purposes, the process of indexing these documents is the primary factor in achieving quality information retrieval and the consequent success of any search mechanism. Gil-Leiva in his paper explains why indexing is contextualized and provides a brief description of some of the most widely used automatic indexing systems [12]. With the above in mind, this work is framed in the lexicographic analysis within the process of an IRS. In this phase, the NER constitutes an independent tool for information extraction, which plays an essential role for a variety of applications related to natural language processing such as information retrieval.

4   Named Entity Recognition

According to [13] the term named entity "...*is a word or sequences of words that are identified as the name of a person, organization, place, date, time, percentage, or amount...*". Therefore, NER aims to recognize and classify such entities in various natural language processing applications. Several works have been reviewed detailing different uses of RE to detect patterns within the text of a document [3, 14, 15]. In the area of NER, a common problem is to obtain relevant information related to some of the mentioned entities, so it becomes important to be able to extract and distinguish this type of elements from the whole set of words that compose a document.

Although some elements are relatively easy to identify by using patterns, e.g., dates or numerical data, there are other elements, such as persons, places, or organizations , that present other difficulties to be identified as belonging to a specific type. In an

IRS, a technique such as NER is very important, as it allows searching for very specific information in collections of documents, extracting and organizing the relevant information. In the work of Sánchez Pérez [13], it is mentioned that in recent years there has been extensive work on the development of NER systems to improve the performance of classifiers using machine learning techniques.

As stated in [1], one of the factors influencing the success of standardization in computing has been the use of RE, a language for specifying text search strings [16].

5   Regular Expressions

Patterns constructed as RE allow the recognition of complexly structured character strings. Their name comes from the mathematical theory on which they are based. In chapter 3 of the book "*Introduction to the theory of automata, languages and computation*" [17], the authors argue that REs can be thought of as a "programming language", in which it is possible to write some important applications, such as text search or replacement applications. In fact, they are recognized by many programming languages, editors, and other tools. Therefore, ERs serve as the input language of many systems that process strings. Examples include the following:

1. Search commands such as the UNIX Grep[2] command or equivalent commands for locating strings in web browsers or text formatting systems. These systems employ an RE-like notation to describe the patterns the user wishes to locate in a file.
2. Lexical analyzer generators, such as Lex or Flex (a lexical analyzer is the component of a compiler that breaks down the source program into logical or syntactic units consisting of one or more characters that have a meaning). Logical or syntactic units include keywords (e.g., while), identifiers (e.g., any letter followed by zero or more letters and/or digits) and signs (e.g., '+' or '<=').

In other words, an RE is an algebraic notation for characterizing a set of strings. They are particularly useful for text search when we have a pattern and a corpus of texts to search. An RE search function will search the corpus and return all texts that match the defined pattern. From the guide in [18] the following example, explained in Table 1, has been developed: an RE can be used to check if an email is valid:

$$|"^[\backslash\backslash w\text{-}]+(\backslash\backslash.[\backslash\backslash w\text{-}]+)*@[A\text{-}Za\text{-}z0\text{-}9]+(\backslash\backslash.[A\text{-}Za\text{-}z0\text{-}9]+)*(\backslash\backslash.[A\text{-}Za\text{-}z]\{2,\})\$" \quad (1)$$

---

[2] https://www.gnu.org/software/grep/.

Table 1. Description of an RE to validate an e-mail.

Expression	Meaning
^	^ at the beginning of the RE, or at the beginning within the []
[\\w-] +	The + symbol indicates that one or more characters must appear within square brackets \\w indicates characters A to Z both upper and lower case, digits 0 to 9 and the symbol _
(\\.[\\w-] +)*	The * indicates that this group may appear 0 or more times. The email may optional-ly include a full stop followed by one or more of the characters in square brackets
@	The following must contain the @ character
[A-Za-z0–9]	In the RE after the @ must contain one or more characters that appear between the square brackets
(\\.[A-Za-z0–9] +)*	Followed (optionally, 0 or more times) by a full stop and 1 or more characters in square brackets
(\\.[A-Za-z] {2,})	Followed by a full stop and at least 2 characters appearing in square brackets
$	Marks the end of an RE

To perform this process, a program written in the C#[3] programming language is used and to solve the issue of RE, the REGEX library is used. C# is a programming language that is included in the .NET Platform and runs on the Common Language Runtime (CLR). The first language of importance for the CLR is C#, much of what is supported by the .NET platform is written in C#. This language is derived from C and C++, is modern, simple, and entirely object-oriented, and simplifies and modernizes C++ in the areas of classes, namespaces, method overloading and exception handling [19].

6 Test Development

This section presents the development of an experimental design test carried out in the context of the research. This aims to study the behavior of REs in the indexing process, in order to identify dates and named entities. The following paragraphs briefly describe the experiment carried out, followed by the methodology and also the parameters used. In order to carry out the tests, a corpus of 497 documents from a public body with similar characteristics, in terms of terminology used to legal documents was used (see Fig. 1).

Each of the documents has a structure similar to the one shown in Fig. 2, where dates in different formats can be distinguished, as well as other named entities. These make up the entries that are intended to be automatically identified and then normalised and incorporated into the repository, and finally used in the process of searching and retrieving the documents.

---

[3] https://docs.microsoft.com/en-us/dotnet/csharp/.

Fig. 1. Representation of the documents used in the experimentation.

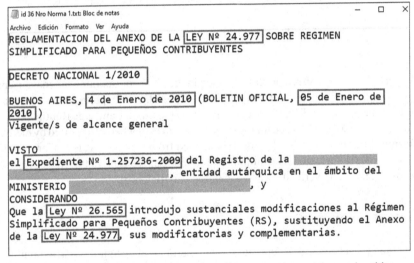

Fig. 2. Fragment of a document showing references to dates and named entities.

In agreement with [3] and also, based on the exploratory analysis carried out, the most common NER pattern is in the following form:

$$\langle \text{Entity Type} \rangle \ [\text{Nro}] \text{ or } [\text{N}°] \ \langle \text{Number} \rangle \ [/\langle \text{Year} \rangle] \qquad (2)$$

where "*Entity Type*" is a part of the named categories. In order to build a table of terms such as the one intended for this work, a common problem is to obtain relevant information related to all the names of the current judicial regulations to be standardised, so it becomes important to be able to extract and distinguish this type of elements from the whole set of words that make up a document.

In Fig. 3, a part of the pseudo code written in C# language and using the Regex library is shown. The following image shows how a breakpoint acts in the Visual Studio debugger. At this point the library recognizes in the selected paragraph, where a named entity is located. This identification is important to be able to normalise these names, and thus, to be able to incorporate them in a pre-established and uniform way.

```
foreach (Match palabra in Regex.Matches(texto, pattern))
{
    if (Regex.IsMatch(palabra.Groups[0].Value, decnorley2))
    {
        if (Regex.IsMatch(palabra.Groups[0].Value, fecha2))  ≤1 ms transcurridos
        {                                    palabra.Groups[0].Value | Q ▾ "decreto 1108/1998\r\n\r\ndel 21/09/1998" ▦▾
            campofecha = palabra.Groups[0].Value.Trim().Split(camposplit,
                    StringSplitOptions.RemoveEmptyEntries);
            tabaux = ControlarFecha(campofecha[2].Trim(), campofecha[4].Trim(), campofecha[6].Trim());
```

Fig. 3. Fragment of the pseudo code to identify named entities.

In relation to date references, which are found within text strings, there are a variety of applications available [20, 21] that help to convert different date formats using RE. Below is a collection of useful RE for finding dates in 'dd/mm/yyyy or yyyy' or 'dd-mm-yyy or yyyy' format:

$$\textbf{RegEx1}: [0\text{-}9]\{1,2\}[0\text{-}9]\{1,2\}[0\text{-}9]\{2,4\} \text{ or} \tag{3}$$

$$\textbf{RegEx2}: \{1,2\}[[1,2\}[0\text{-}9]\{1,2\}[0\text{-}9]\{2,4\} \tag{4}$$

Format 'Month, dd, ', e.g., '4 de julio de 2021'.

$$(\text{Ene}(?:\text{ro})?|\text{Feb}(?:\text{rero})?|\text{Mar}(?:\text{zo})?|\text{Abr}(?:\text{il})?|\text{May}(?:\text{o})?||\text{Jun}(?:\text{io})?|$$
$$\text{Jul}(?:\text{io})?|\text{Agost}(?:\text{o})?|\text{Sep}(?:\text{tiembre})?|\text{Oct}(?:\text{ubre})?| \tag{5}$$
$$\text{Nov}(?:\text{iembre})?|\text{Dic}(?:\text{ciembre}) ?)\backslash s+(\backslash d\{1,2\})\backslash,\backslash s+(\backslash d\{4\})$$

In Fig. 4, another fragment of the pseudo code is shown, in this case to identify a date format within a paragraph belonging to a document, using the Regex library.

```
foreach (Match palabra in Regex.Matches(texto, pattern))
{
    if (Regex.IsMatch(palabra.Groups[0].Value, decnorley4))  ≤ 1 ms transcurridos
    {                              palabra.Groups[0].Value | Q ▾ "4 de enero de 2010" ▦▾
        if (Regex.IsMatch(palabra.Groups[0].Value, fecha2))
```

Fig. 4. Fragment of the pseudo code to identify date format labels with text.

The implemented indexing process can be summarized in the algorithm detailed in Table 2. Line 1 lists the inputs of the algorithm, where: D represents each of the documents to be processed that will be constantly executed, F the date REs, E the named entity REs, M is the term matrix and N is the maximum number of documents to

Table 2. Summary of the algorithm

Algorithm 1: Proposed method to obtain dates and named entities from a set of documents.

1: **Input:** *D, F, E, M y N.*

2: $s \leftarrow 0$;

3: **while** s < M **do**

4:     **for** $p \bullet D_s$ **do**

5:         **if** $p = F$ **then**

6:             *TF ← get_date (p)*;

7:             *NF ← normalize_date (TF)*;

8:         **end if**

9:         *Matriz [s, k] ← valid_if_exists_F(NF)*;

10:        **if** $p = E$ **then**

11:            *TE ← get_entity (p)*;

12:            *NE ← normalize_entity (TF)*;

14:        **end if**

15:        *Matriz[s,k] ← valid_if_exist_E(NT)*;

16:    **end for**

17: **end while**

---

be processed. It is also necessary *p*, for each of the paragraphs composing a document *D* and the intermediate storages are represented by *TF*, *NF*, *TE* and *NE*.

The main loop occurs between lines 3 to 17, where the search and replace of the named dates and entities actually takes place. After being normalized and, after checking that they are only entered once in the matrix of terms, they are incorporated through a function.

7   Results Obtained

Throughout the project, some 6 REs were produced to identify some of the named entities for this experimentation and different date formats. Some of the REs made up covered more than one term to be searched for, as can be seen in Fig. 5.

```
String decnorley3 = "((E|e)xpediente|(E|e)xpte.|(L|l)ey|(D|d)ecreto|(N|n)orma)\\s*" + digito + "+";
```

Fig. 5. Fragment of the pseudo code to create an RE that finds the terms "Expediente", "Ley", "Decreto" or "Norma" (File, Law, Decree or Rule).

The document corpus contains a total of 1,279,205 terms. After running the whole process, it was possible to build the Table of Terms, whose structure is shown in Fig. 6.

The table contains 14,218 terms identified without repetition. In the figure, it can be seen that the program builds the table by assigning a unique ID to each term. Accompanying the term is the number of times it appears in the processed corpus. This value is used by other processes for the weighting of each term. This is not detailed in this paper, but it can be deepened in the presented bibliography [2, 9–11].

```
TablaTerminos.txt: Bloc de notas
Archivo   Edición   Formato   Ver   Ayuda
Clave;Término;Ocurrencia en el corpus
0;reglamentacion;2
1;anex;4
2;ley;6
3;regim;4
4;simplific;4
5;pequeñ;4
6;contribuyent;4
7;decret;2
8;nacional;2
9;1/2010;1
10;air;2
11;4 de enero de 2010;1
12;boletin;2
13;oficial;2
14;05 de enero de 2010;1
```

Fig. 6. Fragment of the Table of Terms with their respective frequencies.

Table 3 presents a summary of the entries found in the processing of the corpus used for the experimentation, including the main named entities detected.

Table 3. Summary of named entities and date formats in the selected corpus.

Reference	Example of the text that appears	Number of times
Art. XX	Art 14	817
arts. XX y XY	Arts. 14 y 17	76
artículo XX	artículo 125	10,404
artículo XX Inciso XXX	artículo 75 inciso 22	853
Ley XXXX o Ley Nº XXX	Ley 26.660 o Ley Nº 24.977	12,650
Expediente XX XXX o expte. XXX	Expediente Nº 1-257236-2009 expte. 20–00160/2000	1,490
Decreto XXX/XXXX	Decreto 1/2010	427

On the other hand, Tables 3 and 4 show the results of the occurrences found in the indexing process, in this case, for date formats, both numeric and text.

Table 4. Summary of the ENs and date formats in the selected corpus

Reference	Number of times	Reference	Number of times
31/12/2015	3058	12/9/69	31
2/12/1964	91	6-11-2009	9
30/4/1970	700	3-1-2001	13
4/2/1971	409	03-10-64	92
31/12/70	24	2-21-94	4
1/02/90	8	20-5-97	1

In Tables 3, 4 and 5 it is possible to visualize the number of different formats that can be found for the same entities in a given corpus. With the algorithm proposed to be implemented, the aim is to bring all these different formats to a uniform one, for example, in the case of dates, to use a format composed of two digits for the day, two digits for the month and four digits for the year.

Table 5. Summary of some of the lettered date formats found.

Reference	Example of the text that appears	Number of times
XX de mes de XXXX	9 de febrero de 2009	1152
XX mes de XXXX	21 octubre de 2005	6
mes –XX	feb-01	48
Mes, XX de XXXX	Abril, 9 de 2007	158

8   Conclusions and Future Work

This paper presents an algorithm for searching and replacing text strings using RE. The performance for incorporating dates and named entities in a term indexing process is analyzed.

Through the experience carried out, it could be proved that a great advantage when applying RE to find named entities is that once the correct expression is defined, and after the corresponding exhaustive search, the entities that match exactly with that pattern will be all the existing ones in the corpus. This is all the more important because legal texts are very structured, and entities appear with a certain regularity.

In turn, the REs are a simple tool to use and do not require more than encoding the expression of the pattern itself, and not training a model for its recognition.

However, a possible disadvantage is their limitation in finding only the predefined patterns, so it is not possible to find another named entity that does not match any of the existing REs.

As is well known, it is of utmost importance to count on the participation of domain experts to validate the terms coming from the corpus, often by carrying out manual checks, which imply a considerable effort.

Thus, in order to reduce the manual intervention effort and to improve the performance in the exhaustive search for patterns, in a next stage, we intend to explore a complementary technique to the RE. This technique is known as Hamming Distance, a similarity metric, which allows us to reduce the dimensionality (number of terms) of the corpus.

As a next step, it is expected to test the algorithm proposed in this work by implementing it in the self-developed IRS. Furthermore, it is planned to build a legal corpus that will be used to evaluate the response times of the IRS and the relevance of the retrieved documents.

Acknowledgment. Thanks are due to the Department of Engineering and Technological Research of the National University of La Matanza, this work is financed within the framework of the PROINCE C241 project.

References

1. Spositto, O., et al.: Propuesta para la construcción de un Corpus Jurídico utilizando Expresiones Regulares. In: 26th Argentine Congress of Computer Science, CACIC 2021, pp. 746–755. National University of Salta, Buenos Aires (2021). http://sedici.unlp.edu.ar/handle/10915/129809. Accessed 25 June 2021
2. Tolosa, G., Bordignon, F.: Introducción a la Recuperación de Información: Conceptos, modelos y algoritmos básicos (2008). http://eprints.rclis.org/12243/1/Introduccion-RI-v9f.pdf. Accessed 25 June 2021
3. Haag, K.: Reconocimiento de entidades nombradas en texto de dominio legal (2009). https://rdu.unc.edu.ar/handle/11086/15323. Accessed 06 Jan 2022
4. Duque Bedoya, E.: Metodología para la Extracción de Metadatos Semánticos de Textos en español utilizando procesamiento de Lenguaje Natural: Subaplicación Para La Identificación De Contextos Espaciales Y Temporales En Textos Que Describan Interacciones Entre Actores. Universidad Eafit Departamento de Informática y Sistemas (2009). https://repository.eafit.edu.co/bitstream/handle/10784/1261/erika_duque_2009.pdf; jsessionid=19D87B68BAFF2D7E3D4296A8C4E727A4?sequence=1. Accessed 06 Jan 2021
5. Rodríguez Inés, P.: El uso de corpus electrónicos para la investigación de terminología jurídica (2008). https://www.tdx.cat/bitstream/handle/10803/286111/pri1de2.pdf?sequence=1. Accessed 06 Jan 2021
6. Cardellino, C., et al.: A low-cost, high-coverage legal named entity (2017). https://hal.archives-ouvertes.fr/hal-01541446/document. Accessed 06 Jan 2021
7. Jurafsky, D., Martin, J.: Speech and language processing (2020). https://web.stanford.e ~jurafsky/slp3/2.pdf. Accessed 06 Jan 2021
8. Robaldo, L., et al.: Compiling regular expressions to extract legal modifications (2012). h www.di.unito.it/~radicion/papers/robaldo12compiling.pdf. Accessed 06 Jan 2021

9. Kuna, H., Rey, M., Martini, E., Solonezen, L., Podkowa, L.: Desarrollo de un Sistema de Recuperación de Información para Publicaciones Científicas del Área de Ciencias de la Computación. Revista Latinoamericana de Ingeniería de Software, 107–114 (2014). http://revistas.unla.edu.ar/software/article/view/81. Accessed 06 Jan 2021

10. González, C.M.: La recuperación de información en el siglo XX. Revisión y aplicación de aspectos de la lingüística cuantitativa y la modeliza-ción matemática de la información (2008). http://www.fuentesmemoria.fahce.unlp.edu.ar/tesis/te.350/te.350.pdf. Accessed 25 June 2021

11. Robredo, J.: Otimização dos processos de indexação dos documentos e de recuperação da informação mediante o uso de instrumentos de controle terminológico. Ciência Da Informação **47**(1) (2019). http://revista.ibict.br/ciinf/article/view/4431. Accessed 25 June 21

12. Gil-Leiva, I.: SISA—automatic indexing system for scientific articles: experiments with location heuristics rules versus TF-IDF rules. Knowl. Organ. **44**, 139–162https://doi.org/10.5771/0943-7444-2017-3-139

13. Sánchez Pérez, C.: Clasificación de Entidades Nombradas utilizando Información Global (2008). https://inaoe.repositorioinstitucional.mx/jspui/bitstream/1009/564/1/SanchezPCR.pdf. Accessed 06 Jan 2022

14. Cucatto, M.: El lenguaje jurídico y su desconexión con el lector especialista: El caso de a mayor abundamiento. Letras de Hoje **48** (1), 127–138 (2013). http://www.memoria.fahce.unlp.edu.ar/art_revistas/pr.9102/pr.9102.pdf. Accessed 06 Jan 2021

15. Dozier, C., Kondadadi, R., Light, M., Vachher, A., Veeramachaneni, S., Wudali, R.: Named entity recognition and resolution in legal text. In: Francesconi, E., Montemagni, S., Peters, W., Tiscornia, D. (eds.) Semantic Processing of Legal Texts. LNCS (LNAI), vol. 6036, pp. 27–43. Springer, Heidelberg (2010). https://doi.org/10.1007/978-3-642-12837-0_2

16. Seghiri, M.: Metodología protocolizada de compilación de un corpus de seguros de viajes: aspectos de diseño y representatividad. Rla. Revista de lingüística teórica y aplicada **49**(2), 13–30 (2011). https://doi.org/10.4067/s0718-48832011000200002. Accessed 06 Jan 2021

17. Hopcroft, J., Motwani, R., Ullman, J.: Introducción a la teoría de autómatas, lenguajes y computación. ISBN: 978-84-7829-088-8, p. 4. PEARSON Ed. S.A., Madrid (2007)

18. Stack Overflow Documentation: Aprendizaje de Expresiones Regulares. https://riptutorial.com/Download/regular-expressions-es.pdf. Accessed 06 Jan 2021

19. Cosio, L., Arrioja, N.: C#: Guía Total del Programador (2010). ISBN 978-987-26013-5-5

20. Regular Expression 101. https://regex101.com. Accessed 06 Jan 2021

21. RegEx Testing. https://www.regextester.com. Accessed 06 Jan 2021

Author Index

Aciti, Claudio   269
Albornoz, Enrique M.   95
Alfonso, Hugo   3
Amadeo, Ana Paola   78
Antonelli, Leandro R.   139
Ares, Fernando Andrés   269
Asteasuain, Fernando   109

Barboza, Lucas   285
Bermudez, Carlos   3
Bianco, Santiago   64
Bossero, Julio César   312
Bourlot, Jimena   95
Buccella, Agustina   124

Cagnina, Leticia Cecilia   170
Caldeira, Luciana Rodriguez   109
Carubia Grieco, Juan I.   285
Cechich, Alejandra   124
Craig, Diego F.   41

Díaz, Javier   78, 285
Dieser, Paula   49

Eberle, Gerónimo   95
Errecalde, Marcelo Luis   170
Esponda, Silvia   297

Fillottrani, Pablo Rubén   237
Fornaroli, Mauro F.   285

Gómez, Sergio Alejandro   237
Gómez, Soledad   78

Harari, Ivana   78

Jorge, Ierache   254

Lanzarini, Laura   64, 159
Ledesma, Viviana Alejandra   312
Lopresti, Mariela   183

Marrero, Luciano   201
Martínez, César   95

Martinez, Victor   159
Matteo, Lorena Romina   312
Méndez, Juan P.   41
Mendoza, Eduardo E.   41
Merkel, Germán   254
Milla, Andrés   21
Minetti, Gabriela   3
Miño, Facundo   285
Moreno, Edgardo Javier   312

Olsowy, Verena   201
Osorio, Alejandra   78
Osycka, Líam   124

Pagnoni, Verónica K.   41
Pasini, Ariel   297
Perino, Franco   219
Pesado, Patricia   201, 297
Pessolani, Pablo   219
Piccoli, Fabiana   183
Presso, Matías   269

Reyes, Nora   183
Ronchetti, Franco   159
Rucci, Enzo   21

Salto, Carolina   3
Santana, Sonia R.   139
Sanz, Cecilia   49
Schiavoni, Alejandra   78
Soligo, Pablo   254
Spandre, Omar   49
Spositto, Osvaldo Mario   312

Taborda, Marcelo   219
Tesone, Fernando   201
Thomas, Pablo   201
Thomas, Pablo J.   139
Torres, Juan Ignacio   297
Tugnarelli, Mónica D.   285

Villegas, María Paula   170

Zangara, Alejandra   64